Inkjet Printing
on Fabric

Inkjet Printing on Fabric

Direct Techniques

Wendy Cotterill

B L O O M S B U R Y
LONDON · NEW DELHI · NEW YORK · SYDNEY

Bloomsbury Visual Arts
an imprint of Bloomsbury Publishing PLC

50 Bedford Square   1385 Broadway
London   New York
WC1B 3DP   NY 10018
UK   USA

www.bloomsbury.com

Bloomsbury is a registered trade mark of Bloomsbury Publishing PLC

Commissioning editor: Agnes Upshall
Copy editor: Jane Anson
Cover design: Eleanor Rose
Page layouts: Susan McIntyre

British Library cataloguing-in-publication data
A catalogue record for this book is available from the British Library

ISBN: PB: 978-1-4081-9190-3

Library of Congress cataloging-in-publication data
Cotterill, Wendy, author.
Inkjet printing on fabric : direct techniques / by Wendy Cotterill.
pages cm
Includes index.
ISBN 978-1-4081-9190-3 (paperback) -- ISBN 978-1-4081-9191-0 (epub) --
ISBN 978-1-4081-9189-7 (epdf) 1. Textile printing. 2. Ink-jet printing. I. Title.
TT852.C68 2014
746.6--dc23
2014015046

Printed and bound in China

Contents

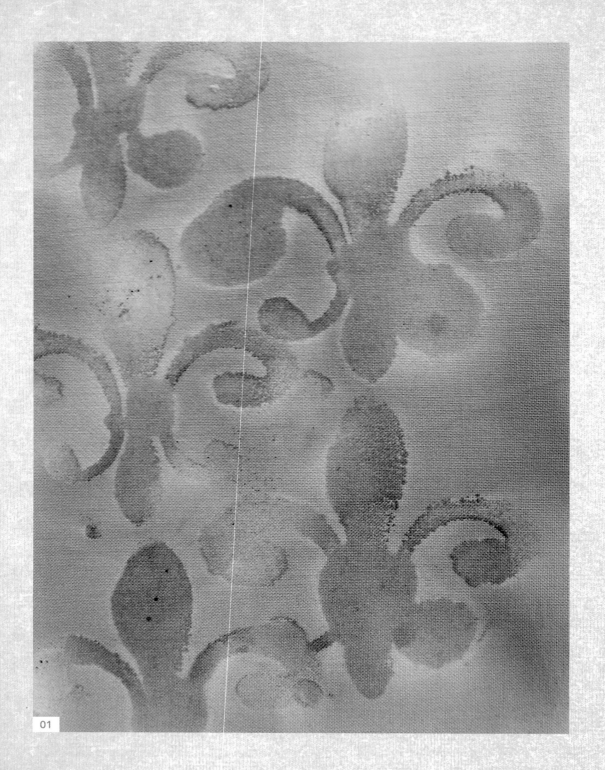

01

Introduction

Students graduating from art and design courses will have encountered many challenges, not least one of resources. After having had access to professional studios and equipment, many have to find alternative facilities or face the enormous expense of setting up a studio from scratch. Many of the techniques outlined in this book have evolved from this sense of frustration, particularly in terms of creating small-scale printed textiles.

Desktop technology – computers, scanners and printers – is now readily available to most of us, and for a relatively small investment it is possible to create printed fabrics for decorative purposes or as prototypes to sample work and process ideas that will subsequently be produced using advanced print techniques. This book aims to look at all the pros and cons of using desktop technology, with plenty of informative guidance along the way. Even mediums that would otherwise be not suitable for printing on fabric, such as water-soluble inks, can be used for creative techniques.

These techniques are a way to bridge this creative gap, and my hope is that fellow stitchers, art quilters or just those with boundless curiosity will find this book useful.

The software used to illustrate techniques in this book is Adobe Photoshop Elements. I use a very old version, but the basics are the same whichever version you use, or for that matter, whichever program you use. The overall design of the interface may vary from version to version, and basic elements may change their name. Moreover, for beginners, the absence of overly complicated functions and over-designed screen elements can help to demystify the process. Other photo-editing programs will for the most part offer the same

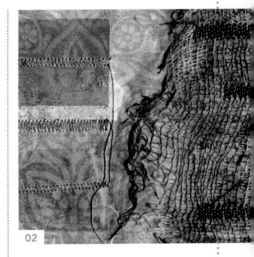

02

01 *A plain piece of cotton that has been printed with a foam stamp and a non-polymer print medium, with a single colour printed over the top and discharged with a small amount of water.*

02 *(detail) This composite piece uses in part the scrim and PVA method described in Chapter 8, but also uses water to discharge some of the background image on paper, as well as additional stitching.*

processes, though the commands may be located in different menus or given different naming conventions, but if you have a good working knowledge of one program, getting up to speed with another should not be problematic. However, using a photo-editing program is not central to this book. The aim is to use desktop printers as a printmaking tool and no, you cannot break your printer by simply placing fabric in it!

The creative techniques illustrated in this book are *direct to fabric* techniques and the book does not examine the use of transfer papers or any other indirect method of printing ink onto cloth. The ideas described may combine Photoshop Elements with a particular print method, or an inkjet print with another simple print process. All the materials and equipment used are easily accessible and most are not specialist equipment – apart from your printer, of course!

Note

Actions carried out in the software program are described in a concise form, for example, the instruction 'In Photoshop Elements, to create a new file, go to the File menu and select New', is written 'File > New'.

All Photoshop Elements screenshots © Adobe Systems Software Ltd.

Digital technology

2

If you are of the generation who did not grow up with a desktop computer, you may not feel comfortable about using one for anything but the simplest of tasks. However, this book aims to look at using the equipment already sitting on your desk and adapting skills you may already have to serve a more creative purpose, particularly for use with fabric.

Desktop computers

Any detailed information here may, to some extent, be out of date before the book even hits the shelves, but any computer bought in recent years will not encounter problems such as insufficient memory or poor processor speed. A modern desktop or laptop computer that can cope with games and online television will have enough capacity for the tasks you require of it.

Unless your computer is very old, it will have more than enough memory and processing capability to deal with moderately high-resolution static images (photographs). As the relative cost of hardware reduces, you may consider purchasing a monitor of 21 inches or more. A screen can get cluttered very quickly and a small screen offers very little space to see your image clearly. An old PC may not have suitable connection sockets for additional equipment such as scanners and printers.

I would recommend a good grounding in basic computer skills. Many people I speak to are easily thwarted by a perception that they have no usable IT skills. Get some good instruction in basic computer skills, even in programs that you may feel are unrelated, such as word-processing.

03

03 Cameras with a macro facility will take very close-up shots.

This will give you confidence and you will acquire some transferable skills such as learning how and where to save your files.

The following website offers a downloadable self-assessment guide to help you assess your own computer skills: www.wendycotterill.co.uk

If you really are allergic to using a computer, then you still have options. See the information on stand-alone or all-in-one printers on pages 12–14.

Tablet computers are not really suitable for using in conjunction with the techniques described in this book.

Cameras

The widespread availability of cameras in mobile phones means that we can now easily capture images on the hoof, but the criteria for choosing a camera are based on the ability to capture good-quality images at a reasonable resolution.

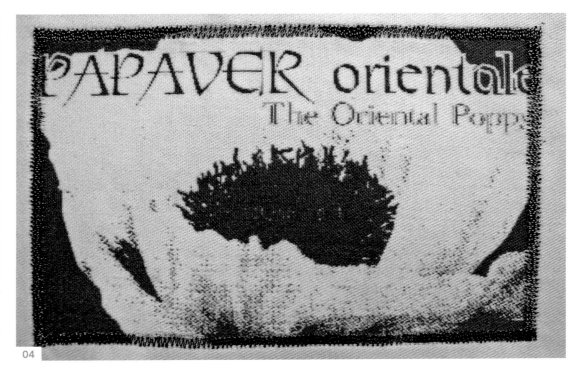

04

04 *If you can get fabric to go through a laser printer, it will result in a very crisp image.*

Phone cameras

The inbuilt cameras on most mobile phones are quick to set up and use, but mostly only capture images of low resolution, and phones do not have lenses to take anything other than 'snapshots'. Nevertheless, photos taken on a phone make a useful *aide-mémoire*.

Digital cameras

Digital cameras fall into one of three categories: compact, bridge and SLR (single lens reflex). Some compact cameras may be no better than a phone camera in their range of capabilities, whereas a bridge camera will be able to capture images at varying resolution, focus, image size and so on. A digital SLR will do the same and more, but will be a larger and heavier piece of kit to carry around. To capture the most versatile images for use in your artwork, I would recommend a compact or bridge (halfway between a compact and an SLR) camera so that you can capture images at a minimum resolution of 300ppi (pixels per inch, see page 25) in a non-compression format such as TIF, RAW or native formats such as PSD (Photoshop format). Familiarising yourself with the technical manual will be enormously rewarding in the long run.

Scanners

There are two kinds of scanning equipment currently on the market.

All-in-one (scanner with inkjet) printer

The beauty of all-in-one printers is that an image can be scanned and printed in one process without the need for a computer. The resulting image is only as the scanner can scan it, but serves as a useful entry level tool.

Note that many all-in-one scanners and printers are wrongly referred to as photocopiers, which causes confusion. An office (laser) photocopier scans and prints an image using an entirely different technology.

Be sure of the difference between the common usage of the terms 'photocopier' and 'all-in-one' scanner/copiers. More sophisticated all-in-one printers may allow some image size and colour adjustments.

Dedicated scanners

A scanner will require scanner software to be installed onto your PC. The Acquire Image dialogue offers various options on which kind of image source to scan, e.g. when scanning from a book you may not notice that the image is made up of microscopic dots. A more sophisticated scanner can select a method of scanning this kind of image that will counteract this effect (descreening). It will also be able to adjust the output, e.g. resolution, colour, size and image quality. Dedicated scanners allow for more pre-scanning image adjustment and can usually scan to higher resolutions and specifications.

Instructions on using a scanner will be discussed in the next chapter with examples.

Printers

One of the most exciting aspects of the digital revolution is the ability to print direct to an incredible range of surfaces. Previously, fabrics needed to be printed using traditional rotary screen printers – which still have their place for creating luxury fabric surfaces – but digital printing enables most of us to use simple printer inks to create immediately available, one-pass printed textiles.

05 *Grayscale image discharged with water and overprinted with additional images plus hand colouring.*

05

06 *Black outline image overprinted onto a gradient-fill background.*

Choosing a printer is very much a matter of personal choice but need not be expensive, as an entry-level printer will produce just as good an image as a more sophisticated printer. The only real criterion for consideration is whether to have a two-cartridge system, which usually uses dye-based (water-soluble) ink, or a four-plus cartridge system that usually uses pigment-based inks. There is no particular advantage to having an A3 printer.

Laser printers

Mono laser printers, particularly A4 size, are now much cheaper than they were, but offer limited possibilities for printing on fabric, not least because the toner has to be heat-set between two closely set rollers at a relatively high temperature. Nevertheless, if your fabric is not heat sensitive and will pass through this kind of printer, you will achieve an image. The resulting image is only sitting on the surface of the cloth and, although water-resistant, will not prove to be particularly robust with time or washing. Colour laser printers can also be used.

Inkjet printers

Most computer owners still use desktop inkjet printers, which are the most versatile tools for creating images on fabric. There are many myths about how difficult it is to pass fabric through a printer, or about the pre-treatment required. Most are untrue and unnecessary, but this subject is discussed later in Chapter 6 Printing on fabric, and advice on ink selection can be found in Chapter 5 Printer inks.

Image-editing software
Photo-editing software (bitmap images)

Adobe Photoshop (printing and photography industry standard)
Adobe Photoshop Elements
PaintShop Pro
CorelDRAW (including a program for vector-based images not normally needed for everyday imaging)
Serif Craft Artist

Drawing software (vector images)

CorelDRAW
Adobe Illustrator

Choosing and using software

Software for creating images is commonly referred to as 'graphics' software, but this is only a general term and you should familiarise yourself with the difference between 'drawing' software and 'photo-editing' software (although many newer programs have been designed to perform both tasks).

Drawing software

Vector-based images are formed of a mathematical outline which reconstructs the image shapes every time the image file is opened. Drawing software is used in very specific circumstances for designing commercially printed fabrics so will not be addressed in detail in this book. Images created in this format are images, with limited numbers of blocks of colour. The advantage of vector-based images is that resolution is not an issue, as the image will be just as crisp whether it is printed at 1 × 1 inch or 1000 × 1000 inches. The disadvantage is that they can only accommodate around 256 colours, and photos can use up to 16.7 million colours.

Photo-editing software

Photo-editing software is used for editing photos that are pixel based. If you zoom in on a digital photo, the image becomes a series of coloured pixels. The larger the dimensions and resolution of an image, the more computer processing speed and memory is required.

Photo-editing software will not test the capability of most newer PCs, and is available at a reasonable cost. As with printers, having the most up-to-date, industry-standard software is not necessary, and can be avoided either by using older versions of high-end software or exploring the many smaller companies now producing custom software for artists and crafters. They all follow pretty much the same principles and concepts, even using the same terminology for many effects, e.g. cut, copy and paste commands are the same in most software.

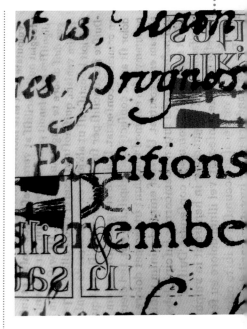

07 *Random text and outline images.*

08 *Two grayscale images, one printed onto voile and the other onto plain cotton, overlaid.*

15

As outlined above, image files are constructed either as bitmaps or vectors. Photo images are bitmaps. This can be demonstrated by opening a photographic image on screen and zooming in on the picture, you will see little squares (pixels) appearing. When you zoom back out, the image reforms because the pixels dither to form a clear overall image. There are many drawing and photo-editing software packages on the market. Most work in similar ways, but will have differences that you may not expect. For example, Adobe Photoshop is primarily a photo-editing package and, although it has been tweaked over the years to accommodate some vector-based features, it was not developed as a drawing package. CorelDRAW was developed as a vector-based drawing package but has been enhanced to provide more photo-editing features.

If you are able, try different software before buying. You can do this by downloading a trial version from the software company's website. Most trial versions are fully-functioning pieces of software with the 'save' function disabled. You may find one that is much more intuitive to use. I use a very old version of Adobe Photoshop Elements for almost all of the image editing I carry out, partly because it was the one I learnt to use first, but also because it offers 80% of the functionality of industry-standard Photoshop at around 20% of the cost. Photoshop is a piece of software that, although possibly off-putting to beginners, will pay back dividends as it has the flexibility to do custom effects. Older computers used not to have the capacity to install photo editing software, but this is no longer the case. If you can get some good basic instruction, your confidence will grow. You do not need to be an expert, but you do need to be able to navigate your way around your program of choice.

09 *Sometimes the back is more interesting than the front!*

09

Image basics

Understanding file formats

Before taking photographs or scanning images, it's useful to know a little about file formats, as this will help you to select the appropriate equipment and the right settings on your camera or scanner.

Use a camera that offers the ability to save images at 300ppi and in a file format that does not compress the image when saved, such as .TIF or .PSD. Many popular cameras may have fewer options and will only save images in a .JPG format, which is a compression (lossy) format. Compression formats were developed to streamline the amount of file information contained within the image file and prevent them from occupying too much disk space. .JPGs do this very efficiently, but every time a .JPG file is resaved, the file is re-compressed, and one effect of this is to reduce the quality of the image so that more and more detail and clarity will be lost. Therefore, if your camera offers the option of using a non-compression format such as .TIF, consider this option as a format to save all your original images. They can be converted to .JPG later. Typically, phone cameras will only save low-resolution .JPGs.

10 *This is a full-colour image scanned at 300ppi. The small yellow square in this image is enlarged as a detail in the next image.*

11 *This is the enlarged section of the original scan and demonstrates what a photographic image looks like when enlarged: the pixels become visible.*

10

11

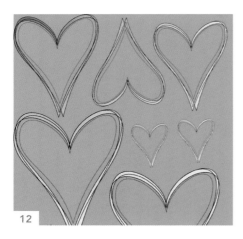 

When opening an image to edit it, always save the edited image as a new file. DO NOT overwrite (save over) the original image.

Scanners are operated from scanner software installed on your computer and will offer various file-format options to save images.

Other sizing issues are looked at in more detail in the next chapter, but remember that no amount of reformatting or resizing will correct a badly taken photograph.

Choosing and using cameras

The rule of thumb, as with most technical equipment, is 'the more you pay the more you get'. If your image is photographed on a phone camera it will probably be limited in dimension, resolution and so on. You will need to investigate the options on your camera and decide if you need a different one to capture the kind of images you want. The following will serve as a useful checklist.

Do you want to take images in great close up?
You need a macro lens facility. Do not confuse this with a zoom lens facility.

Do you want your images to be captured in a non-compression format (i.e. not .JPGs)?
Again, you will need to investigate the possibilities of a camera that offers greater functionality. Capturing an image as a .JPG and resaving

12 *This is a .JPG image that has been created with no resolution or dimension alterations.*

13 *This image demonstrates what happens when a .JPG image becomes over-compressed or if the image is enlarged too much. The smudgy background is caused by a problem known as compression artefacts.*

it into another file format is not the same as capturing the original in a non-compression format, although you may not mind or may not notice the difference. The choice is yours.

Do you want higher-resolution (300ppi) images?
You will probably need to upgrade to a compact or bridge camera. Capturing an image at 72ppi and upsizing it to 300ppi is definitely not a good idea – you will notice the difference and it won't look good!

Do you want a degree of manual control over focus and depth of field?
As a minimum you will need a bridge camera or may want to upgrade to a digital SLR camera. Using a digital SLR camera is exactly the same as using a manual SLR.

Camera settings

With any new piece of equipment, familiarising yourself with all its capabilities and functions and how to set them pays dividends in the long run. This is particularly true of a digital camera. The following is a list of settings to consider:

- Set the image size settings (physical dimensions) to one of the preset sizes your camera offers, e.g. 800 × 600 pixels.

- Set the resolution to 300ppi and save the images in a non-compression format such as .TIF or .PSD wherever possible.

- Select any of the preset options that will affect the overall quality of your image, such as lighting preferences, default dimensions, resolution and so on. It is worth spending a little time investigating your options, particularly those relating to dimension, resolution and file format.

- Upload your images from your camera to your PC regularly, and use a logical folder structure and a consistent naming regime.

Choosing and using scanners

In essence there are currently two varieties of scanner to suit all budgets. An all-in-one scanner-printer will enable you to scan and print images with or without being connected to a PC. A dedicated scanner will give you the option of capturing images from various sources including film negative, slides and printed sources.

When opening the scanner software on your PC, the Acquire Image dialogue box will open and offer various options on which kind of image source to scan, as well as the opportunity to adjust the output e.g. resolution, colour, size and image quality. Dedicated scanners allow for more pre-scanning image adjustments and can usually scan to much higher resolutions.

Open the Scanning dialogue box and set all of the appropriate parameters: resolution, colour, output purpose (e.g. website), dimensions etc.

If you are scanning an original image, set the resolution to at least 300ppi, in full colour and at the original dimensions. Any editing can be carried out on scanned images. Follow the same filing routine as for photographed images.

Selecting and setting scanning options in the Scan dialogue box in the scanning software can be counterintuitive, particularly when scanning grayscale or black and white images, as it is easy to assume that a black and white image holds less information for the scanner to detect. This is true in part, but as images are made up of pixels, in order for an image containing fine outlines to be recreated accurately, pixels need to be created on a much smaller scale, i.e. higher resolution. Most outline-type images are scanned at 600ppi.

Scanning tips

Full-colour images

Scanning an original at 300ppi will result in an accurately detailed copy of the original. If the colour in an image prints out differently from the way it is represented on the screen, your monitor probably needs adjusting or calibrating rather than the colours in your image.

Grayscale and black and white

The terms grayscale and black and white are often used interchangeably; for instance a photograph with no colour is often called 'black and white', although in reality it is grayscale.

14 *Detailed images printed in black only will produce a much clearer image.*

Black and white

The definition of black and white for these purposes is quite different in that the only information in the image is black, with the white being provided by the print surface. This results in very high-contrast images which easily lend themselves to creative techniques such as those demonstrated in Chapter 7 Using colour, where instructions are given for converting colour images into either grayscale or black and white.

The two images below demonstrate the difference between grayscale and black and white.

15

16

15 Coneflower *A colour photograph of a simple flower head has first been isolated from its background and then converted to grayscale. Image > Mode > Grayscale.*

16 Coneflower 2 *The same image has subsequently been converted into a black and white (bitmap mode) image. Image > Mode > Bitmap.*

Grayscale

Grayscale is what it says on the tin – black at one end of the tonal range, white at the other and every percentage of grey in between. Do not confuse grayscale with black and white.

If you have a pencil drawing with lots of shaded areas, there is no need to take up unnecessary space on your hard drive by scanning it in full colour. Set your scanner settings to capture the image in grayscale mode, but at an *absolute minimum* of 300ppi. Finely drawn lines will break up at 72ppi. The image can then be cleaned up by converting it to black only.

If you are scanning an image that is mainly made up of fine lines, such as a pen and ink drawing, scan the original at 600ppi in grayscale mode. Any less and the image will break up. You may then choose to convert the image to black and white (only).

17 *Grayscale images can be used as background texture in an image that only has a spot of jewel colour.*

18 *Grayscale images contain all of the detail in a full colour image, but the subject matter becomes clearer, enabling it to be seen distinctly through an additional layer of semi-transparent fabric.*

Basic image adjustments

4

Sizing images

The most common and useful tasks you will need to carry out on your images are to do with sizing, but sizing adjustments can take different forms and require a basic understanding of how an image file can be resized.

When referring to an image the term 'size' is usually meant to refer to its physical dimensions when printed. If you have photographed your image with a camera, the physical dimensions will have been allocated via the camera settings.

However, when moving into the digital domain, physical dimensions become relative. Your camera settings will allocate size in terms of a pixel measurement, for example 480 × 640 pixels. This in turn is relative to the resolution. If an image's physical dimensions are 480 × 640 pixels at a resolution of 72ppi (pixels per inch), then 480 × 640 pixels will print an image at approximately 6.6 × 8.8 inches, i.e. 480 ÷ 72 = 6.6 and 640 ÷ 72 = 8.8. If the resolution is set at 300ppi, then 480 × 640 pixels will print out an image at approximately 1.6 × 2.13 inches, i.e. 480 ÷ 300 = 1.6 and 640 ÷ 300 = 2.13. The best general guidance here is that if you have enough capacity on your camera memory card, set the resolution at 300ppi and the dimensions as large as possible – up to 2592 × 1944 pixels. You may need to purchase a new memory card with a larger capacity.

To discover the dimensions of an existing image in Photoshop Elements: Image > Resize > Image Size.

Two dialogue boxes here demonstrate the difference between two images; both have physical dimensions of an A4 print, but they are set at two very different resolutions. Note the difference in how much disk space these two files will occupy by referring to the information given at the top of each box under Pixel Dimensions.

Dimension

Dimension is what is meant when talking about the size of an image. This will be the size of an image when it is printed.

Resolution

Resolution refers to the density of pixels contained within an image and measured in pixels per inch. Again, ensuring that the original image is captured at the correct resolution will avoid the need for unnecessary changes which will result in image degradation.

Resizing an Image

When the dimensions of an image are either increased or decreased, the total number of pixels can either be added or subtracted to accommodate the change in size, or the total number of pixels can remain the same. Increasing or decreasing the total number of pixels utilises a process called resampling. Both increasing and decreasing the number of pixels in an image will result in image degradation, so limiting the amount of change to the dimensions of image to around 10 per cent will limit the degradation. It is for this reason that accurate camera or scanner settings will limit the need for resizing.

To alter the dimensions and the number of pixels of an image, open the image and select Image > Resize > Image Resize. With the Resample Image and Constrain Proportions boxes selected, change the Document Size settings. Click OK.

If the dimensions of an image are increased without increasing the number of pixels, the image quality will result in degradation. If the dimensions of an image are decreased without altering the total number of pixels, the image quality will be increased, as each pixel will have been made smaller to fit into the decreased dimensions of the image.

To alter the dimensions of a file whilst retaining the same number of pixels within the image, open the image and select Image > Resize > Image. Deselect the Resample and Constrain Proportions check boxes and alter the Document Size measurements. Click OK.

Altering Resolution

To change the resolution of an image, with the image open, select Image > Resize > Image resize. With the Resample Image and Constrain Proportions check boxes selected, change the resolution settings.

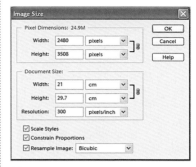

19

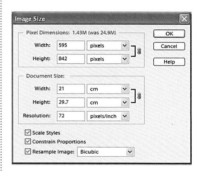

20

19, 20 *These two Image Size dialogue boxes show the differences in the number of pixels required to render an A4 image at 72ppi and 300ppi.*

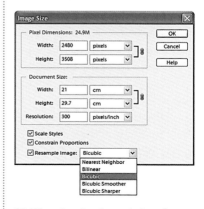

21 *When changing the resolution of an image, there are several methods of resampling.*

Constrain Proportions

If the dimensions of an image need to be changed without altering the resolution or the resolution needs to changed without altering the dimensions, this check box should remain selected.

Why is resolution important?

If you want to use an image in a commercially printed book or publication, you will need to supply the file with a minimum resolution of 300ppi or the image will look 'foggy'. However, if you are publishing your images on a website, reduce the resolution to 72ppi as a matter of course, as a computer screen is unable to display at a resolution higher than this, so you will be using up precious server space if they are saved at 300ppi, and your images will load very slowly. In addition, photographic images for the web need to be in a specific file format, currently .JPG, or less commonly .GIF or .PNG.

Resampling

Resampling is a simple concept, though not always easily understood. When a photo-editing program has to increase the size of an image while maintaining the same resolution, it has to increase the number of pixels within the existing space. The program has to work out what colour these new pixels need to be, based on one of a number of mathematical formulae. Each resampling method will render a slightly different result, depending on the original image.

Similarly, when decreasing an image size by reducing the number of pixels in an image, the program has to combine existing pixels and work out which colour to render a new pixel that has been created by combining one or more existing pixels (compression). If you are mixing paint, certain colours will produce a predictable result, e.g. one part red and three parts white will produce pink; but in resampling a digital image, depending upon the method used, one red and

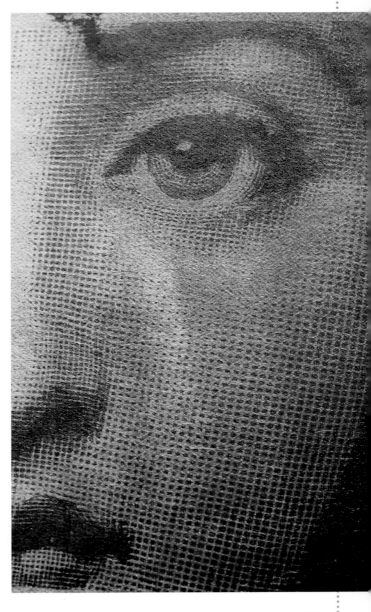

22 To add an extra dimension to a black outline image, it has been printed on fabric that has a gradual variation in colour.

three white pixels will produce a range of different results, but more than likely three white and one red pixel will result in a new white pixel (as white is the predominant colour). This effect seen in isolation is not such a problem, but if this method is employed across a whole image, many of the red pixels will be deleted, with the visual effect of making the image more white in certain areas and therefore less crisp. So different resampling methods will deliver varying results, depending on the original image, and no resampling method will produce perfect results.

In summary, changing the dimensions of an image as well as resampling will result in a degree of image degradation. Taking a photograph at the desired dimensions and resolution in the first instance will save you a lot of time in the long run and will result in a sharper image.

Cropping

Cropping is a technique that cuts out a piece of the image – as if you were cutting off the edges with a pair of scissors. The image quality will remain unaltered.

Cropping tool

If you want to cut out parts of an image, the Crop tool will do the job perfectly. Select the Crop tool from the Tools palette. Place the cursor over the image and click and drag a box around the section you want to cut out. Press the enter key to finish the crop. To undo the crop, Edit > Undo and start again.

23 *Crop tool.*

24 *This is an example of a black and white image.*

25 *A cropped section from the centre of the previous image.*

24

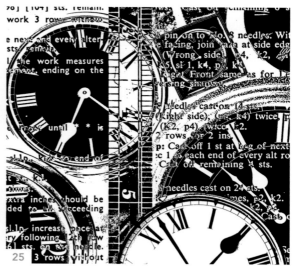

25

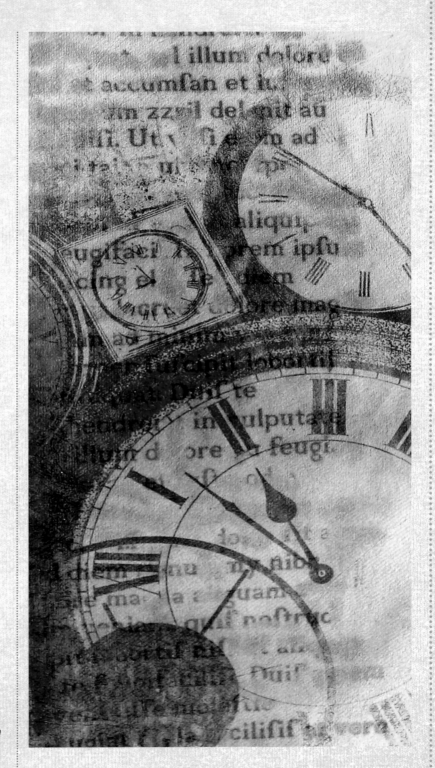

26 *This piece of fabric was printed using a cropped section of the image file shown on the left.*

Printer inks

5

A brief introduction to printers was given in Chapter 2, which outlined that desktop printers are either inkjet or laser (toner) printers. Each type has its own range of possibilities. The key issue to be discussed regarding printing direct to fabric is the type of ink available for use in desktop inkjet printers.

Dye-based inks

Dye-based inks are water-based. They are not permanent, will run in water, are prone to fading in daylight (UV light) and for the most part would not be the ink of choice. However, do not dismiss them because of that, as this solubility will enable you to use discharge techniques, similar to other textile discharge-dyeing techniques.

Two-cartridge print system

The two-cartridge inkjet printer accepts two cartridges: one black and one tricolour. This is the simplest and oldest kind of printer available, and almost always uses dye-based ink. However, this ink offers the most flexibility and versatility for creating many of the images shown in this book, particularly when used in conjunction with Print.Ability print medium (see page 39).

Pigment inks

At first glance, pigment inks seem to be the best choice for printing textiles. They are largely UV resistant and can be used on all fibres. Unlike dyes, pigments do not dissolve in water and are only held in

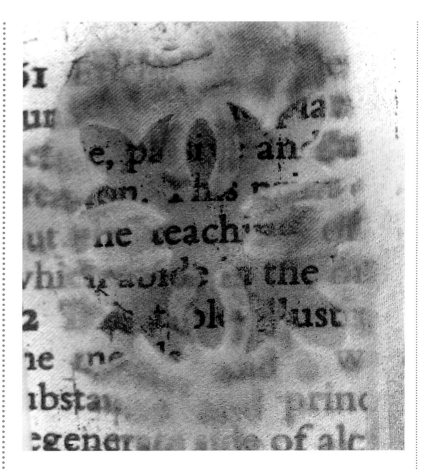

27 *Evolon is a particularly absorbent fabric; in conjunction with the print medium Print.Ability and water to discharge the background print, the text in this case bleeds, but the stamped image does not (see Chapter 6 Printing on Fabric).*

suspension, so a pigment-based ink is not absorbed into the fibres but only sits on the surface of the fabric. Pigment inks must contain a resin to 'glue' the pigments onto the fibre. The addition of this resin limits the amount of pigment that can be included in the ink.

Four (plus) cartridge system

The four/six/eight-cartridge system accepts black, cyan, magenta, yellow, pale cyan, pale magenta, etc., all in separate cartridges. This is a system that has been developed to improve the results for photographic printing. These printers tend to be more expensive to purchase and maintain. The fact that pigment ink does not discharge into water does not mean that it will withstand machine washing or repeated handwashing. Some artists recommend that you do not heat-set pigment-based inks, as the bonding of the ink to the fabric is compromised.

28 *Black (mixed) ink printed on nylon voile and spattered with a small amount of water. When the water dries, tide marks are left around the edge of the ink.*

Dye versus pigment inks

The discussion of dye vs. pigment can become very technically involved, so fear not, all you need to know is whether the ink in your printer will discharge (run) in water. Test this simply by taking any waste printout from your printer and immersing it in water. Try both a black and full-colour print, as you will probably find that the ink in the black cartridge is more or less permanent. The coloured printout will also indicate which of the three colours (cyan, magenta and yellow) discharges more easily. You might also like to try the same test with an old print and heat-set it with an iron to make comparisons.

N.B. All laser and inkjet printers can be used in combination with other transfer methods, such as T-shirt transfer papers etc., but these are largely offset processes, which are not the subject of this book.

There are now printers that use permanent dye-based ink as well as sublimation (transfer) dye, acid dye and fibre-reactive (Procion) dyes developed for specialist printing. However, you cannot change the type of ink your printer uses, and printers that use specialist inks require a non-thermal delivery method in order to produce successful prints. Although these inks have potential for use in desktop printers, they are manufactured for semi-industrial wide-format printers, are very costly and the dyes require steaming to fix them. A brief explanation of these types of printer is given in Chapter 13 Taking things further.

Black ink

Black ink in most desktop printers is an archival (more or less permanent) black ink and is less likely to discharge, but again if this is taken into account when developing creative methods, you can create some fantastic results.

Some two-cartridge print systems can operate with just a tricolour cartridge being present, as a form of black can be mixed from cyan, magenta and yellow. This is useful if you want your black ink to discharge.

Ink bleed on fabric

There are different ways of preventing ink from bleeding on fabric, the most popular one being that of changing the print quality to 'Draft'. This option appears in the Print dialogue box. However, mostly this will just create a very pale print and image quality will be lost.

The best method to use is to temporarily reduce the resolution of your image. If your image is set at 300ppi, change the resolution to 72ppi, print the image and then undo the edit by going to Edit > Undo or just close the file without saving.

Reducing resolution prevents too much ink being deposited on the fabric surface. If an image is set at 300ppi, the printer will deposit a spot of ink for every pixel. Temporarily setting the resolution to 72ppi deposits less ink, but distributes it more evenly across the surface of the fabric.

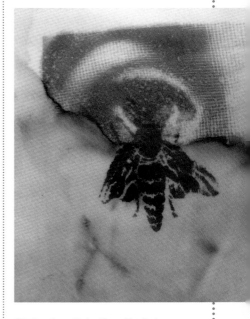

29 *Dye-based ink will readily discharge into water, but this can be used as a design feature, even if it was a happy accident.*

30 *This image has been printed onto silk habutai with the default setting selected for the amount of ink normally required to print onto paper.*

Running costs

Empty cartridges can be refilled from a bottle of refill ink, but it does get a little messy. Always wear thin latex gloves for this task, and once the cartridge is full, place it back in the printer and leave overnight if possible, to allow the ink to settle. Be careful not to touch the electronic contact points on the front of the cartridge. Occasionally the ink printer heads will block, so gently wipe them (usually underneath the cartridge), with a clean fibreless cloth. You will eventually need to replace the cartridge, but you can reasonably expect to get a minimum of six refills before this happens, so the cost per print reduces dramatically.

The potential for using desktop printers as a creative tool is detailed in Chapter 7 Using colour and Chapter 8 Creative techniques, where a desktop printer is used as a tool to create images on fabric but is only the starting point in the creative process. Refill ink is usually dye-based, so is a cost-effective means of producing a lot of experimental discharge prints or images.

31 *Water discharging as a considered method can be controlled if used in conjunction with a non-polymer print medium.*

32 *A simple background colour has been discharged with a water and mica powder mix to highlight the discharge effects.*

Printing on fabric

6

Fabric for inkjet printing

The question I am most frequently asked is, 'How can I print on fabric?' The simple answer is: 'Put the fabric through the printer and see what happens.' At the risk of stating the obvious, many people imagine hurdles that just aren't there.

Most smooth, light- to medium-weight fabrics are capable of passing through a printer, although each printer has its own idiosyncrasies. Below is a brief guide to using fabrics in a printer.

Natural fibres

Most natural fibres have the ability to absorb moisture and will absorb printer ink readily. Some, such as silk, have a tendency to over-absorb inks, depending on how tightly spun or woven the fabric is; for example, silk habutai will cause bleed lines. This issue of over-inked fabric is discussed on page 33.

Synthetic fibres

Single-source synthetic fibres (with the exception of nylon) are not able to absorb moisture, and any ink will sit on the surface of the fibre and dry without being absorbed. This makes the dried image appear much paler.

Mixed-fibre fabric such as polycotton is processed to create a textured yarn and the small fibres have the ability to absorb some of the ink.

> ▶ **MATERIALS CHECKLIST**
> - Plain woven fabric
> - A4 labels
> - Desktop A4 inkjet printer
> - Print.Ability (optional)
> - Scourer e.g. Metapex (optional)

33 *Image printed onto scrim, which has a very open-weave structure.*

Preparing to print

The only requirement to enable fabric to travel through an inkjet printer is to make the fabric rigid, thereby imitating the nature of paper. This is done by supporting the fabric on a sticky label.

Printer feed system

It is nearly always assumed that a top-feeding printer will cause fewer problems than one that feeds paper from underneath, as the pathway through the printer demands less bending. However, in practice there is little or no difference.

Older, very basic printers sometimes have a means of manually adjusting the rollers to allow slightly thicker fabrics to pass through.

This can sometimes be achieved by looking at the options in your Print dialogue box and selecting an option for thicker paper or card. Sometimes this can be achieved by selecting glossy print paper. Smooth fabric, even if it is more of a medium weight, will always travel more easily through a printer.

Open-weave fabric will appear not to have created a very successful print – whether on natural or synthetic fibres. In practice, however, fabrics such as organza or muslin will receive as much ink as any other fabric, but the gaps in the weave will not allow your brain to close the visual gaps to complete the image.

Slubbed fibres will not support a well-defined image. If you can get the fabric to pass through a printer at all, the resulting image will be broken up by the rough surface but nevertheless will produce interesting results.

In conclusion, preparing fabric to enable it to travel through a printer is straightforward, and if done meticulously you will avoid 90% of problems.

Preparing the fabric

- Cut your fabric slightly larger than A4.

- Lay it on a flat surface (press with an iron first if necessary).

- Remove a single A4 sticky label from its backing and carefully lay it down on the fabric.

- Gently rub on the fabric side to remove any air bubbles or creases. Creases are the biggest single problem that will cause your printer to stop printing. It is not hard to remove the fabric and reset the printer, but you are likely to waste ink or printed pieces in the process.

- Make sure there are no threads hanging from the cut fabric (though you will be very unlucky if this causes a problem).

- Clip a tiny triangle of fabric from the two corners of the leading edge. Mostly this isn't necessary, but it is irritating if it causes a misprint. The sandwich of fabric and sticky label creates two layers that are put under tension as they feed through the printer. As the fabric sandwich reaches the exit rollers, you can just observe that the two corners are tipping up slightly. It is just possible that

34 *Lay the fabric down on a flat surface and place the sticky label down on top of the fabric.*

35 *Clip a small amount of fabric from the corners of the leading edge of the fabric to printed. Fabric does not need to be routinely pre-treated with any kind of printing medium before passing through an inkjet printer. If the ink you are using is normal printer ink, it will be more or less absorbed into the fabric fibres.*

this may trigger a misprint if the ink cartridge, which is travelling backwards and forwards across the fabric surface, catches one or other corner. Trimming away a tiny amount of fabric will prevent 99% of these occurrences.

Pre-treating issues

Myth No. 1: You must buy 'inkjet printable fabric' to print through your desktop printer

The 'inkjet printable fabric' available commercially may or may not have been pre-treated. It has probably only been mounted on backing paper to enable it to travel through the printer. Any fabric sold as 'digital fabric' or similar will have been prepared for receiving inks normally used in wide-format printers, such as fibre-reactive inks. This fabric has been pre-treated to bond with the fibre-reactive dye and then be made permanent by steaming the fabric, which has no relevance to inkjet printing. See the information listed in the suppliers section at the end of this book.

Myth No. 2: You need to pre-treat your fabric with a digital medium

Inkjet 'digital printing mediums' marketed by paint manufacturers are polymer-based mediums, not much different from acrylic paint, which are formulated to coat artists' canvases in order to prevent too much ink being absorbed by the raw canvas. This is mainly applied to stiffen the canvas to allow it to pass through the printer unmounted. Painting this medium onto 'normal' fabric will be of little benefit to a textile artist. The printed fabric will still not be washable and the fabric will stiffen significantly. Washing issues are not addressed by the manufacturers of these products, as they are classified as fine-art mediums.

Bubble Jet Set is a non-polymer print medium formulated for inkjet printing and is applied directly to fabric. You can expect to be able to handwash the fabric occasionally in cool water.

Print.Ability is another non-polymer print medium for stabilising inkjet ink on fabric. It can be used to solve printing issues on difficult fabrics that will not absorb ink, such as Lutradur. Again, you may be able to occasionally hand rinse fabric treated this way in cool water, and in some instances a cool rinse will clear the fabric's surface of any loose ink and surplus medium. Print.Ability does not make your ink permanent or washable in the normal sense, but its versatility is that it can also be used to create faux discharge effects on fabric, which are illustrated right and on page 43

36 *A colour-filled background can be overprinted with water. If a small amount of water is used, a tide-mark effect will be created, giving additional texture.*

Myth No. 3: Pre-treatment will make the fabric washable

There is no practical means of printing images with a desktop inkjet printer using regular printer ink that will withstand any kind of regular washing, especially machine washing, not least because the washing process causes the fabric surface to be abraded. The only guaranteed 'washable' images are those applied using transfer papers – commonly referred to as 't-shirt transfer papers'.

Myth No. 4: Using fabric instead of paper will break a printer

You cannot break an inkjet printer just by passing fabric through it. If it becomes jammed, press the eject button, or gently ease it out through the rollers. But please note, you are voiding your manufacturer's warranty if you use substrates that are not recommended by the printer manufacturer.

Print mediums

If you do choose to use a print medium, then whichever medium you choose, there is a case to be made for gently rinsing your printed fabric in cool water to remove any loose ink, and adding a gentle scourer such as Synthrapol or Metapex to prevent back-staining by any loose ink. Do not rub the fabric while rinsing, as this will degrade the image, as will repeated machine washing. Be sure to do test prints before embarking on any project.

Having established that non-polymer mediums do not leave a residue on the fabric surface and that they create a surface on which the ink can produce a crisp, stable image, there are a number of creative techniques that take advantage of these qualities used in conjunction with printer ink.

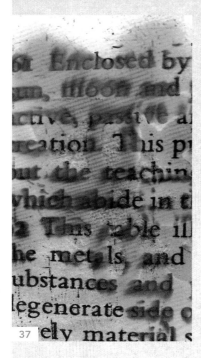

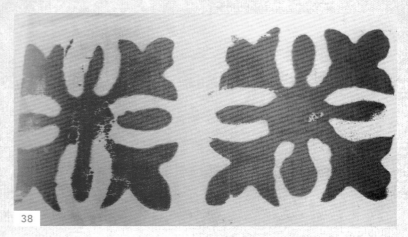

37

38

37 *Bright white cotton will enhance any colour printed or painted on its surface.*

38 *This effect is created very simply. A gradient image file is created and printed on a piece of fabric that has been stamped with a simple shape using Print. Ability as the medium. The finished print is then gently rinsed in cool water and the stamped shape reveals itself as the water discharges the background ink.*

39 *This example was produced by creating a gradient fill and printing it onto a piece of fabric that had a stencil, made from an A4 sticky label, placed on it before printing. Make sure the label is well adhered to the fabric before printing.*

39

40 *This piece takes advantage of the very absorbent nature of Evolon. When water is used as a discharging agent, the fabric will keep absorbing for a long period.*

41 *This effect is created very simply. A gradient image file is created and printed on a piece of fabric that has been stamped with a simple shape using Print.Ability as the medium. The finished print is then gently rinsed in cool water and the stamped shape reveals itself as the water discharges the background ink.*

Using colour

7

Printing full-colour images on fabric will for the most part be disappointing, not least because these kinds of images will probably contain a lot of fine detail that will be lost when printing onto fabric.

This chapter takes a look at removing colour and re-colouring with flat colours. Most of the images we capture with a camera or scanner will be full-colour images. Using images creatively is the way to get the most from them, but this may require a fresh perspective. Try to look at your images less in terms of their subject matter and more from an abstract point of view, taking into account things such as shape, line, texture and contrast. Line, shape and tone become more apparent when colour is removed, but a few preparatory steps are needed before we can do this.

The difference between a grayscale and a black and white image was discussed in Chapter 3, but instructions on how to convert an image to grayscale or black and white and back again are given here.

42 *Full-colour scan of a printed design.*

43 *Scan of image converted to grayscale.*

44 *Grayscale image converted to bitmap using Filter > Sketch > Stamp.*

42

43

44

Removing colour

Most photo-editing software is capable of converting a colour image to a black-only image, but this is usually a two-stage process, the first stage being conversion to grayscale:

Image > Mode > Grayscale

The next stage is conversion to black-only:

Image > Mode > Bitmap (keeping all default options)

If the resulting image is too dark, or too light, or loses definition, keep undoing back to the grayscale image and start again, each time changing the options listed in the Bitmap dialogue box. Every image is different, even though they may look the same. Better images will be produced with a little practice.

There are other ways of creating a black-only image using filters. For example, when the Sketch option in the Filter menu is applied, the image will be automatically converted into a single colour. If the Colour palette is set at the black and white default option, the image will turn

45 *This image, although simple, contains complex detail and has therefore benefited from being converted to a black-only image.*

46 *Again, this simple image has benefited from being converted to a black-only image and seen as a close-up detail. The contrast creates interest.*

45

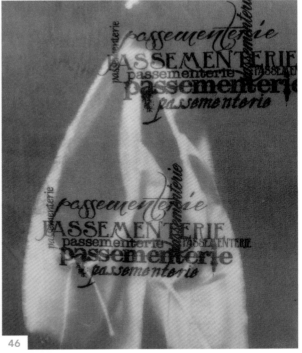
46

47

47 *Dense areas of detail are highlighted by printing on a single-colour fabric.*

directly to a black bitmap image. Other ways of creating a single-colour image are described later in this chapter.

At this point, if when you convert an image the results are disappointing in tonal range, consider adjusting the image for brightness and contrast etc., as this will influence how usable a black and white image you get. Consider altering the options in the Image > Adjust menu.

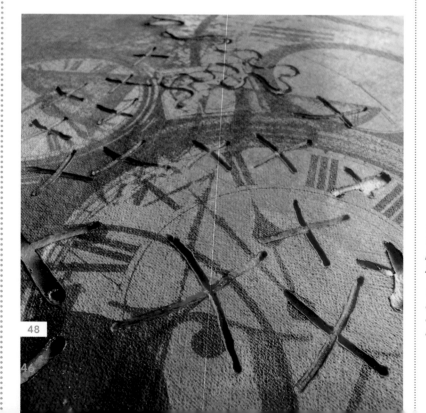

48

48 *This image contains areas of more and less detail, and has benefited from being printed onto a gradient-dyed piece of fabric.*

49 *Using black-only images and overprinting onto a full-colour background creates detail without overly confusing the image.*

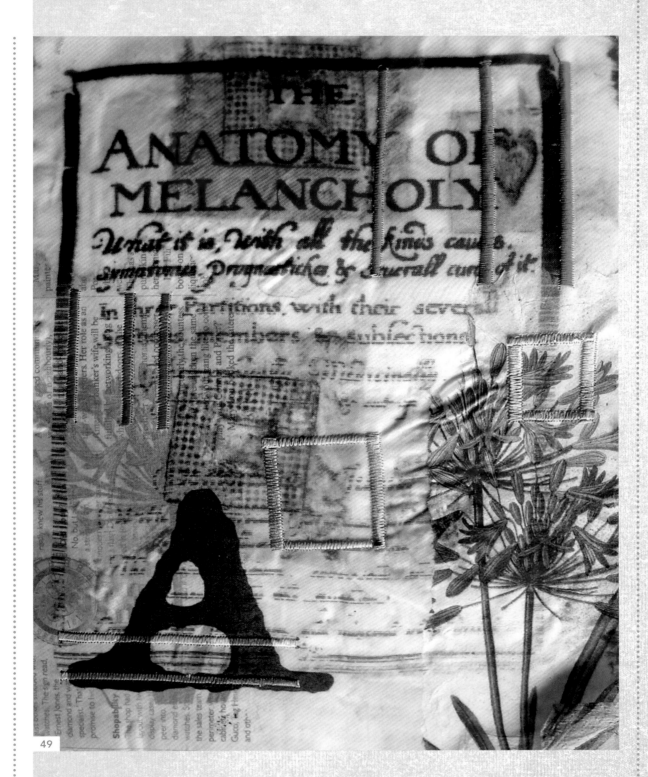

Re-colouring images

An image on a computer screen is created by using a combination of Red, Green and Blue light (RGB). A printed image uses a combination of Cyan, Magenta, Yellow and Black inks (CMYK) delivered in one pass from the printer cartridges.

Full-colour images lose much of their definition when printed on fabric, even white fabric, so consider re-colouring to some degree. Black or medium to dark images work well on coloured fabric and re-coloured images work best on white fabric.

Follow the instructions below to create a single-colour flood-filled image and print on fabric.

Remember that printer ink is transparent and its colour when printed will depend on the colour of the fabric.

Re-applying colour

Converting full-colour images to grayscale or black and white was discussed on page 45. Both types of images tend to have high contrast and create bold images. These images can be re-processed and re-coloured.

A digital image converted to bitmap mode only supports black, so it will need to be converted back to an RGB image, even though it will still only be black at this point:

> Image > Mode > Grayscale
> Image > Mode > RGB

Adding colour with no tonal range (flat colour)

To apply a single flat colour to a single colour image, select the Magic Wand tool from the Tools palette, usually found on the left of the screen.

50 *Rather than using black as a single colour, this image was re-coloured to dark blue and combined with its complementary colour to create impact.*

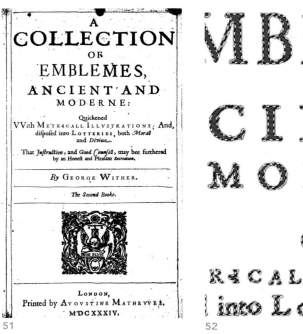

51

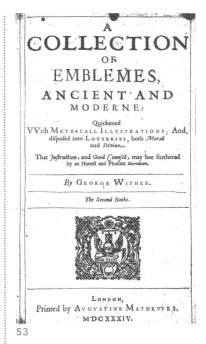

53

52

Once the Magic Wand tool is selected, click into the Colour Palette at the bottom of the Toolbox and a dialog box will open to enable the selection of the correct foreground colour. When the correct colour is selected, click Okay.

Next, click on a black area, and all the black areas of your image will be selected. In the Options palette above the open file, make sure that the 'Contiguous' tick box is deselected.

| 54 | ✱ | ▣▣▣▣ | Tolerance: 32 | ☑ Anti-aliased | ☐ Contiguous | ☐ Use All Layers |

With the black image still selected, go to Edit > Fill Layer. In the Fill Layer dialogue box, drop down the Use option and select Foreground Colour. Click OK.

Your image is now filled with the new colour.

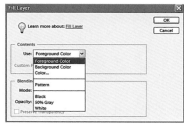

55

51 *Image file has been converted to a single black colour but making sure it is in RGB mode.*

52 *The same file when selected with the Magic Wand tool. If the Contiguous option is deselected in the options bar, all the black areas of the image will be selected.*

53 *While the black areas remain selected, go to Edit > Fill Layers and select the Foreground Colour to apply the colour in the Colour palette. Click OK.*

54 *Whenever a tool from the Tools palette is selected, an Options palette will open immediately above the file space, with different options to customise the tool capabilities.*

55 *Photoshop offers many ways of selecting a fill colour, which is designed to improve the workflow, but it is also possible to select colours as you go.*

56   57

56 *Grayscale image converted back to RGB colour mode.*

57 *Grayscale image with Colourize option applied in the Hue/Saturation dialogue box.*

Adding colour with tonal range

A simple way of adding a single colour to a grayscale image whilst maintaining the tonal range in a grayscale image is as follows: with the image selected, open the Hue/Saturation dialogue box; Image > Adjust > Hue/Saturation. Tick the Colourize box and move any or all of the sliders in any combination, which will apply a colour to the image.

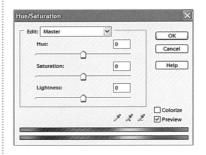

58 *Tick the Colourize box. By moving the sliders, a single colour will be applied to the image. If no colour is applied, check to see that your image has been converted back to full-colour mode. Image > Mode > RGB Colour.*

Colour shifts

You may not always want or be able to convert your image into a single-colour image – some of the continuous tone contained within the image is integral to the image quality. Most photo-editing programs will give you the option to change all the colours in an image simultaneously in the form of a colour shift, that is, all the colours move a step around the colour wheel.

Open a full-colour image.

Enhance > Adjust Colour > Adjust Hue/Saturation as before, but this time move the sliders without ticking the Colourize box.

59

59 *The original image as scanned.*

60

60 *A small colour shift has been applied by using the Hue/Saturation dialogue box, which will let you preview the effect as you move the sliders.*

Investigating the Hue/Saturation dialogue box will give all kinds of combinations of results for you to explore. For instance, rather than applying a colour shift to all the colours in the image at the same time, select one colour and apply a change individually.

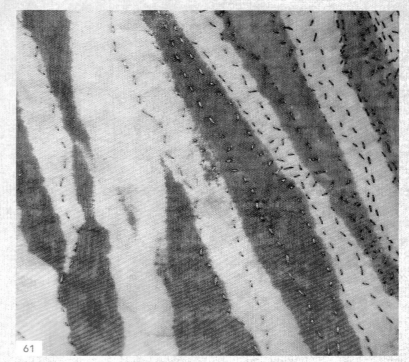

61 *Bold mark-making can be softened with a gradient fill.*

62 *The text on this piece was printed onto printed and stitched fabric.*

Digital colour

Single-colour fills

We sometimes overlook the fact that colour is an element in its own right. One of the most useful features in a photo-editing software package is the ability to create an image that consists entirely of colour, i.e. a colour fill. In this instance the instructions describe how to create a single-colour flood fill and a gradient (two or more colour) fill.

Selecting colours

The current foreground colour appears in the upper colour selection box in the toolbox; the current background colour appears in the lower box.

63 *Printing a piece of fabric with a single colour on an inkjet printer can give unexpected results. If the print is slightly uneven, this will be accommodated more easily, as the eye will accept the uneven nature of the print better than if it were printed on paper.*

64 *Use the advantage of dye-based ink to discharge (remove) colour.*

A: Default Colours icon
To restore the default foreground and background colours, click the Default Colours icon in the toolbox.

B: Switch Colours icon
To reverse the foreground and background colours, click the Switch Colours icon in the toolbox.

C: Foreground Colour box
To change the foreground colour, click the upper colour selection box in the toolbox, and then choose a colour from the Colour Picker palette.

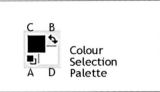

65 *Colour-selection tools.*

D: Background Colour box

To change the background colour, click the lower colour selection box in the toolbox, and then choose a colour from the Colour Picker palette.

Create a simple colour fill using the Paint Bucket tool:

> File > New
> Set up an A4 document at a resolution of 72ppi.
> Choose a foreground colour.
> Select the Paint Bucket tool.

Click the part of the image you want to fill. All specified pixels within the specified tolerance are filled with the foreground colour or pattern. Change colour by repeating this process.

Print this image on a piece of white fabric and you will have created a piece of coloured fabric. If you print this colour fill onto a pale-coloured fabric you will obtain a third colour as printing inks are transparent, e.g. yellow ink on pale blue fabric will produce green.

Gradient fills

Two-colour gradient

The Gradient tool creates a gradual blend between two or more colours. You can choose from preset gradient fills or create your own.

To create a gradient colour fill:

> File > New
> Set up an A4 document at a resolution of 72ppi.
> Select the Gradient tool.
> In the options bar, choose a fill from the gradient sample preview box.
> Click the triangle next to the sample to view a preset gradient fill.
> Click inside the sample to select a preset gradient fill.

This tool works slightly differently from other tools: instead of just clicking in the file, you need to click and drag where you want the gradient to appear. Try clicking and dragging from edge to edge to start with, but the gradient can be drawn from inside or outside the file. Select the linear (straight line) gradient, which will shade from the starting point to the end point in a straight line.

66 *Paint Bucket tool.*

67 *To choose a new colour, place the cursor over the colour bar to the right of the large Colour palette. This will select the appropriate area of colour ranges you want. The large colour selector area on the left will change to accommodate your new selection and you can now make a precise colour selection by clicking the mouse within this area. Click OK when you are done.*

You can change the gradient's shape. In the Gradient toolbar, which appears when you select the Gradient tool, there are different gradient shapes, e.g. radial gradient. All are applied with the same click-and-drag motion.

69

Pre-set
Gradient options

GRADIENTS
Linear, Radial, Angle
Reflected, Diamond

69 *Gradient tool Options palette.*

If you only want to fill a shape with colour, select the Marquee tool, click and drag a shape and, while it is still active, the Gradient tool will apply the gradient to this shaped area only.

If you want to create your own gradient, reselect the Gradient palette, select the New Gradient option and create your own colour options by selecting each of the tabs that automatically populate the Gradient box. As you select each tab, the Colour palette will open. After creating your new gradient you will able to save it with a new name. It will be available every time you reopen the file.

If you are colour-filling a single file, as many of the previous examples did, once a colour has flooded the file, you can change the colours in situ:

Enhance > Adjust Colour > Adjust Hue/Saturation

The sliders will change the hue (colour), and saturation/lightness (tints/tones).

70 *This is a black-only image that has been selected with the Magic Wand from the Tools palette, making sure that the Contiguous check box in the options palette has been deselected.*

71 *With the black flower still selected, select the Gradient tool and click and drag the cursor over the image. When the mouse is released the gradient will only fill the selected area.*

Creative techniques with water discharge

By employing some of these colour-fill techniques and printing them straight onto fabric with dye-based ink, you will have created surfaces ready to be discharged easily with water or water-based pastes, simply overprinted with fabric paint or black outline images. The term 'discharge' in terms of dyeing cloth is just a process whereby colour is removed from fabric. This can be in the form of powerful bleaching agents, or in this case, by simply using water to remove water-soluble inks. Discharging is not possible using pigment-based inks.

72 Fabric printed with a single-colour fill creates a background for any number of other overprinting techniques. The example has been overprinted with a black outline image as well as fabric paint printed using a foam stamp.

73 Similar to the previous image, just a different colour background ink.

72

73

Creative techniques

8

Not all fabrics are suitable for printing in an inkjet printer, either because the fibres may not be absorbent, such as synthetic organzas, or because the fabric is very loosely woven, such as cotton scrim. I have developed two techniques that enable the use of either of these fabrics.

Scrim on paper

This method requires several hours' drying time.

Cut a piece of soft, undyed cotton scrim to A4 size and lay it flat on top of a piece of A4 copier paper. Copier paper is tougher than many people expect, and is especially strong when wet.

Place a dessert spoon (20ml) of undiluted full-strength PVA glue in a small pot and gradually dilute with approximately 100ml of water. You may need to do tests, as fabric cannot be removed from its backing paper if the PVA glue has not been sufficiently diluted.

Treating fabrics with a synthetic medium such as PVA would normally be avoided as it substantially alters the handle of fabric, but in this case, very dilute PVA used to treat scrim will impart just enough rigidity to the fibres so that the scrim is more easily handled than in its untreated state.

Method

With the scrim laid as flat as possible, paint the diluted PVA through the scrim onto the paper with a flat brush. You will be able to manipulate and straighten the scrim with the brush to some extent once it is wet. Allow the scrim/paper sandwich to dry completely.

> ▶ **MATERIALS CHECKLIST**
> - Synthetic organza
> - Cotton scrim
> - A4 copier paper
> - Household emulsion paint
> - PVA glue
> - Desktop A4 inkjet printer

Once dry, trim the edges and loose threads as you would for the sticky-label printing method (see page 38) and print on it using your inkjet printer. (Some printers may not like this substrate, so be prepared.) Once the image is dry, gently but firmly remove the scrim from its backing.

The beauty of this technique is that you get two images from one print! As well as the printed scrim you will also get the paper backing with a variation on the image and unexpected effects.

First, you may get an impression of the weave of the scrim on the backing paper, producing a faux fabric effect. You may also get a roughened surface resulting from the dried PVA, which can make the paper's surface feel like fabric.

If you place the two parts of the sandwich back together you can create a kind of optical illusion when observed from a side angle. The fabric and the paper are simply placed back together but the fabric is no longer in the same position on the paper. This illusion is difficult to capture in a photograph as it relies on three-dimensional vision, but suffice to say that you are seeing the image on the scrim and the image on the background in very close proximity and the eye creates the illusion.

74 This technique is simple to create and relies only on basic ingredients that need to be prepared accurately, following the instructions given.

75 You need to peel the two layers apart very carefully, however copier paper is stronger than you might expect. The weave of the fabric has had the effect of a stencil and created a 'printed' weave effect on the base paper layer.

74

75

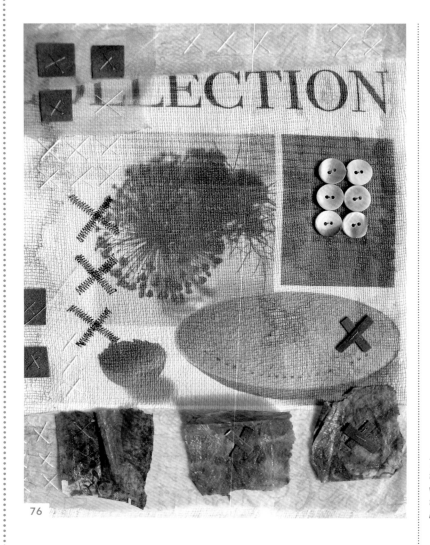

76

76 This piece of work was created using the scrim and PVA method but the layers were not split apart. The paper gives body to enable embellishments to be stitched into the piece.

Organza and paint

The previous technique can be replicated, this time using synthetic organza and household emulsion paint. The paint needs to be diluted in exactly the same way as the PVA glue above, and painted onto a sandwich of synthetic organza and paper, but the dilution of the emulsion paint, unlike the PVA method, is not critical. Print the organza paper sandwich as before.

Again, coating fabric with any kind of medium will alter the handle and thickness of an otherwise sheer fabric, but suitably diluted

77 Small pieces of coloured organza have been glued onto paper with diluted PVA and printed with faces. Some have been left unprinted. The pieces have been assembled into a group of nine.

77

emulsion paint will impart a very light coating onto the organza, which in turn will leave a residue absorbent enough to capture sufficient ink to form an image which, if printed onto uncoated synthetic organza, would dry to a very pale result.

If you want to push your printer even further, you can try printing over the top of fabric that has already been machine stitched. Printing over hand stitching is not a good idea as it almost bound to catch on the printer cartridge.

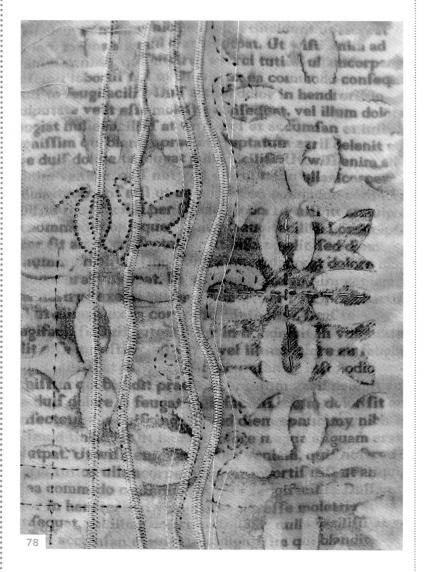

78 Dyed Evolon overprinted with black text, printed with Print.Ability solution and over-sprayed with dye-based colours. The highlighted areas are created by using a craft soldering iron to etch into the surface of the Evolon, enhancing any applied colour.

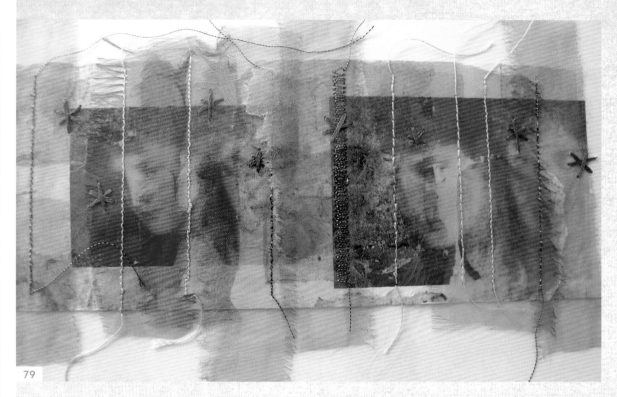

79

79 *A full-colour image has been overprinted on synthetic organza and tissue paper glued to paper with additional hand and machine stitching. By varying the colour of the paint and the organza, this technique offers even more possibilities, with each slight variation producing subtle colour shifts.*

80 *All of these pieces have been printed on white or natural fabric and embellished with outline stitching.*

80

81

Image effects
using filters

9

Most, if not all, photo-editing software includes filters. A filter is simply a sequence of effects that has been pre-programmed within the software and can be applied to an image or part of an image in one click.

Some filters are almost imperceptible to the eye and are meant to make minor adjustments such as sharpening or blurring an image, or are designed to be used in conjunction with another effect. However, remember that no filter effect will correct a badly taken photograph. For example, inexperienced users often mistake the Sharpen filter for an effect that will correct an out-of-focus image.

Some filters are designed to work with an image containing colour, and some will automatically render an image into a single colour, usually the colour that is designated as the foreground colour in the Colour palette. If the Colour palette is set to its default option of black and white, the image will be converted to black and white.

Applying filters

This technique is in essence an all-in-one process, as the filters are pre-defined effects for you to apply to your images.

With a file open, click on the Filter menu on the top menu bar, and try out as many as you want in order to gain knowledge of how each filter option works.

81 *The centre panel of this large collage was created by printing text onto a piece of plain Evolon. A piece of coloured synthetic organza was then laid over the top and cut, using a metal edge and a craft soldering iron, to form the square images that melt onto the background Evolon.*

Method

To use the Filter Gallery, choose Filter > Filter Gallery, select a category, and click the filter you want to apply.

Choose a submenu, followed by the filter you want to apply. If a filter name is followed by ellipses (…), a Filter Options dialogue box appears.

If a dialogue box appears, enter values or select options.

Applying filters can consist of many steps and in later versions of Photoshop Elements there is a facility to record all the steps applied to an image in the History palette so that the effect can be recorded and saved for later use on another image. Filters can be applied to single colour fills as well as complex images.

All the following filter effects have been created by applying a filter effect to a gradient fill (see Chapter 7 Using colour).

Any of these simply created effects can be printed onto fabric and used as a coloured background.

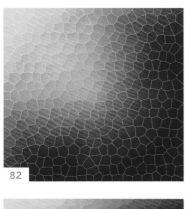

82

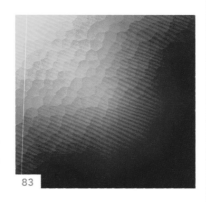

83

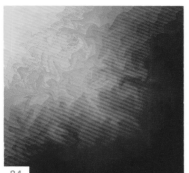

84

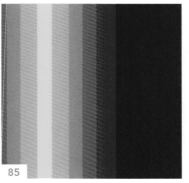

85

86

82 *Filter > Stained Glass.*

83 *Filter > Stained Glass plus Dry Brush.*

84 *Filter > Stained Glass plus Dry Brush plus Liquefy.*

85 *Filter > Artistic plus Poster Edges.*

86 *Filter > Noise plus Ink Outlines.*

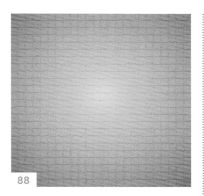

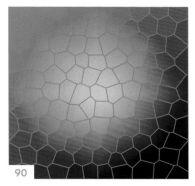

87 *Filter > Artistic plus Sponge.*

88 *Filter > Mosaic Tiles.*

89 *Filter > Mosaic Tiles.*

90 *Filter > Mosaic Tiles.*

Filter options

The following are descriptions of a selection of filters that are most useful for creating special effects for images to be printed on fabric.

Adjustments filters

Change the brightness values, colour, grayscale range and tonal levels of pixels in an image. Convert colour pixels to black and white.

Artistic filters

Simulate a painterly appearance on traditional media and create a unique look.

Brush stroke filters

Give a painterly or fine-art look using different brush- and ink-stroke effects.

Distort filters

Geometrically distort an image, creating three-dimensional or other reshaping effects.

Pixellate filters

Sharply define an image or selection by clumping pixels of similar colour values.

Render filters

Create 3D shapes, cloud patterns, refraction patterns and simulated light reflections.

Sketch filters

Add texture for a 3D effect or hand-drawn look.

Stylize filters

Produce a painted or impressionistic effect by displacing pixels and heightening contrast.

Texture filters

Give the appearance of depth or substance, or add an organic look.

All of the effects listed above may or may not create any discernible effect depending on what type of image they are applied to.

91 *One single image has been processed in Photoshop Elements and nine variations have been printed on one layer of muslin.*

91

Using text in images

10

The one aspect that continues to attract the most interest is using text as a design element to be printed onto fabric. Putting words in a picture is a curious device and much has been written on the subject. Placing a recognisable word in a place where pictures and colours live is a double-edged sword. It helps the audience to engage with the piece of work, but at the same time the audience bring their own experience to that word and its meaning, which was not necessarily what was intended.

However, the concern here is how to get text into your image. Many of the photo-editing programs now treat text separately from other aspects of an image. Photoshop and Photoshop Elements treat text as a vector outline, in much the same way that vector programs do. Each piece of continuous text is placed in its own layer, which means that when any piece of text is reselected it is still editable.

It is only when applying a filter or other creative effect that the text will be converted to a bitmap image – that is, an image that is no longer editable.

- Create a new file.

- Select the Type tool in the Tools palette and click in the file space.

- Type your words and you will see a new layer automatically created to accommodate the text.

- Click away from the text and the type is deselected.
- If you want to change the font, reselect the text and open the font options dialog box, in the options bar immediately above the picture file. Other options are also available, such as changing the size, formatting, leading, colour and other special effects.

Another way of altering the size of the text is to deselect the text, deselect the Type tool and reselect the text with the Move tool. The text is now selected like an image, which enables you to move the body of the text across the picture. You can also resize it by dragging the corners, but be careful not to alter its proportions, as it can look very ugly.

With the text selected with the Move tool, the Filter menu now becomes available and most, if not all, filters can be selected. It will become obvious though that if the text is not 'fat' enough any filter effect applied will be hard to see. When selecting a filter a warning message will appear on the screen, stating that the text will be 'simplified'. This means that the text will be rasterised – converted to a bitmap – and will no longer be editable. Click OK. If each piece of text is kept on its own layer, it is easy to delete that layer and create another one.

Sometimes, if you are using text as a design feature, it may be an idea to place each word on its own layer as the formatting in some programs can be limited and moving each word to be placed is a quicker or better option. Once all the words are placed relative to each other, link all the type layers together in the Layer palette, by selecting one type layer and putting a tick in each subsequent layer link box to the left of the layer. This means that when you select one word, all the others in the link chain will be selected and can be moved in one action. These separate layers can later be merged into a single layer.

One of the best treatments for text is to maintain the integrity of its appearance and shape and to apply other effects. In the section dealing with blending modes (see page 80), simple colour-combining effects can be created in only one or two steps, so if text is treated as an image, place it over the top of an image in the Layer palette and apply a blend from the layer options bar, and you can get as many different options again if you apply a different colour to the text.

Another effect which is very popular, but I recommend using judiciously, is that of filling letter forms with an image. This can be done in many ways, but here I describe one of the simplest methods.

- Create a new A4 file.

- Type the words you want to use in a bold font in black.

- Layer > Flatten Image. The text is now part of the image background and will no longer be editable.

- Click on the background layer in the Layer palette and convert the locked background into a layer.

- Select the black text with the Magic Wand tool. All of the text will be selected; then simply press the backspace key.

- Open a second image and drag the picture into the picture space of the text file, using the mouse. The image is now placed on a layer but will be covering the text. Drag the picture layer below the text. This method enables the text and picture layers to remain separate, a feature not always possible in older versions of software.

Other simple effects

While the text is selected with the Magic Wand tool you can also apply a fill or a gradient fill. Try using the pre-set Metallic effects in the Gradient palette.

92

92 *Simple black text forms the basis of a piece of work that has had pieces of synthetic organza fused to its surface using a craft soldering iron.*

Rather than deleting the colour from the text, apply a Transparency effect in the Layers palette.

Try the same effect, but this time with a colour other than black as the text colour.

If you want to create a piece of text that is a piece of calligraphy or your own handwriting, the process is similar. Instead of creating 'live' text, scan the handwriting with the scanner set to 600ppi and at grayscale to be converted to black only after scanning. This is normally a more versatile way than scanning in black only, as you have more options that create final black-only text.

Open the handwritten text file, then select the black parts of the image with the Magic Wand and follow the same instructions as for other images.

93 Simple lines of text are useful to form a background that has interest but is not sufficiently distracting to interfere with another image placed over it.

94 The text printed onto the background here has been given a focus using just a small embellishment.

93

94

Digital collage

11

Layers

Layers are like stacked, transparent sheets of acetate on which you can place all the separate elements of an image. You can see through the transparent areas of a layer to the layers below. The bottom layer in the Layers palette, labelled 'Background', is created as a locked layer. You cannot change its stacking order, blending mode, or opacity unless you convert it a regular layer. You can work on each layer independently, experimenting to create the effect you want. Each layer remains independent until you combine or merge the layers.

Adding layers to an image increases the file size of that image.

Layers are useful because they let you add components to the image and work on them one at a time, without permanently changing your original image. For each layer, you can adjust colour and brightness, apply special effects, reposition layer content, specify opacity and blending values, and so on. Use the Layers palette and Layer menu in the editor to manipulate your layers.

This example demonstrates the process of adding a gradient colour as a layer above a grayscale image. If you open a grayscale image and create a gradient, the gradient fill will be placed in its own new layer. Look in the Layers palette and you will see an option to alter the opacity of the colour fill. Change the opacity of this layer to gradually reveal the image underneath.

You may have noticed by now that when any element, whether image or text, is placed in the file space, it can be picked up and moved with the Move tool, but unlike a word-processed document, the edges of the

95

96

95, 96 *Before and after: a grayscale image has been overlaid with a semi-transparent gradient fill.*

picture will disappear as it moves beyond the edges of the file space. If the file is saved as a .PSD or .TIF, all the layers remain available and all the elements remain moveable and editable.

Collage techniques

Collage is a long-established art technique that involves combining images with glue, paint or any other method. A photo-editing program will allow you to do this digitally and then you can print out the collaged image, in this case, on fabric. You might also consider building up a collage manually, by printing each image separately.

If your printer will allow it, try also printing on a piece that has already been collaged. Older printers are particularly accommodating if there is a manual lever, usually located on the underside of the printer, to adjust the printer rollers and increase the space for the fabric to pass through.

Combining images using layers

Combining two or more images from separate files is very straight-forward. If the image being moved has a transparent background, transparency will be maintained when it is moved.

97

97 Damask. *No result using this technique can ever be predicted. Newer printers are much less likely to achieve this.*

Select two images, making sure that both image files have approximately the same physical dimensions and resolution.

Open both images in your photo-editing software. Click on the main image to activate it. Click and hold the left mouse button and drag the first image over the top of the second. As you drag you will see an outline surrounding the first image. Release the mouse when you are hovering over the second image file.

The two images will now be placed on separate layers, one on top of the other. If your image is not in the correct place you will be able to click on it and drag it to another part of the background. If you have created more than one layer you can move a layer by simply clicking and dragging the layer further up in the Layers palette.

If an image on the uppermost layer does not have any transparent areas, you will not be able to see any part of any images on the layers below. One visual effect to employ here is to blend the two layers.

There are a couple of simple one-step moves that help to blend layers, by decreasing the transparency of a layer or by blending one or more layers.

98 *The inkjet-printed text image forms the basis of a fabric collage that relies on a fusing process using synthetic fabrics.*

Blending layers

Adjusting opacity

Much like filters, a description of the Blend effect is meaningless seen in isolation, but is worth experimenting with as it is a one-click effect.

The result of blending layers is entirely dependent upon the images being blended, and some blends may not appear to make any difference.

A layer's opacity determines the degree to which it obscures or reveals the layer beneath it. A layer with 1% opacity is nearly transparent, and a layer with 100% opacity is completely opaque. Transparent areas remain transparent regardless of the opacity setting.

You can use layer-blending modes to determine how a layer blends with the pixels in layers beneath it. Using blending modes, you can create a variety of special effects.

Simple blending effects

The process of combining two images is very simple. Place the cursor in the image you want to be the uppermost image, then click and drag it to the second image. Close the first image file.

At first you will only be able to see the uppermost image, but if you change the Blend Mode option and in this case choose Lighten, the resulting mix is instant.

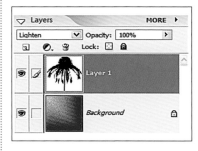

99 *Layers palette showing where the two elements are placed within the image file.*

100 *The uppermost layer contains the black image on a white background.*

101 *This is the resulting image when the uppermost image is blended with the underlying gradient fill using the Lighten option.*

102 *This is the result when the same black image is blended with the underlying gradient fill using the Difference option*

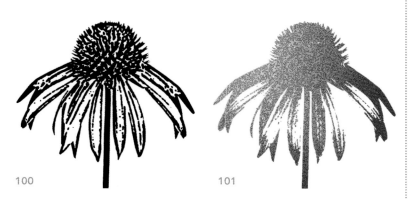

100

101

102

103

104

105

103, 104, 105 *All images use an image file that has been constructed in layers so that an effect can be created separately in each image.*

All the examples here are digital collages using layers and blending modes as listed below and printed directly on fabric.

Blend-mode options

Not all blend effects produce a discernible effect and are entirely dependent upon the structure of the images that you are using. Below are some that you may find useful, together with a brief technical explanation.

Normal

Edits or paints each pixel to make it the chosen colour. This is the default mode.

Darken

Looks at the RGB colour information separately and selects the base or blend colour – whichever is darker – as the result colour. Pixels lighter than the blend colour are replaced, and pixels darker than the blend colour do not change.

Colour burn

Looks at the RGB colour information separately and darkens the base colour to reflect the blend colour.

Linear burn

Looks at the RGB colour information separately and darkens the base colour to reflect the blend colour by decreasing the brightness.

Colour dodge

Looks at the RGB colour information separately and brightens the base colour to reflect the blend colour. Blending with black produces no change.

Linear dodge

Looks at the RGB colour information separately and brightens the base colour to reflect the blend colour by increasing the brightness. Blending with black produces no change.

106

106 *Unlike the previous image, a digital image was created as a single layer and printed on fine silk habutai. This image was created by collaging pieces of fabric, ribbon and printouts on a page of a notebook, which was then digitally scanned. Blending effects were subsequently applied to the digital image.*

Soft light

Darkens or lightens the colours, depending on the blend colour. The effect is similar to shining a diffused spotlight on the image.

Hard light

Multiplies or screens the colours, depending on the blend colour. The effect is similar to shining a harsh spotlight on the image.

Vivid light

Burns or dodges the colours by increasing or decreasing the contrast, depending on the blend colour.

Linear light

Burns or dodges the colours by decreasing or increasing the brightness, depending on the blend colour.

Pin light

Replaces the colours, depending on the underblend colour. This mode is useful for adding special effects to an image.

Hard mix

Reduces colours to white, black, red, green, blue, yellow, cyan and magenta – depending on the base colour and the blend colour.

Difference

Looks at the colour information in each channel and subtracts either the blend colour from the base colour or the base colour from the blend colour.

Exclusion

Creates an effect similar to but lower in contrast than the Difference mode.

Pattern repeats

12

Whole books have been written on creating digital patterns and pattern repeats, but this is a keep-it-simple approach, so here are two methods that can be used in Photoshop Elements, and better still, saved within the program for use another time.

Random repeats

The simplest method for creating repeatable elements in a background can be found in the Brushes palette.

A digital brush is very good way of creating reusable shapes. Defining a brush is much the same process as creating a pattern.

Open a simple image file to begin with so that the process is more easily understood, preferably a black-only image. Select which part of the image you want to copy – all of it, or parts of it – using the Magic Wand tool. Once selected, Edit > Define Brush from Selection and name the brush. The image is now created as a pre-defined brush in the Brushes palette.

To apply, open a new file, select the Brush tool from the Tools palette. The Brush option bar will open above the picture space where your new brush will be placed at the end of the pre-set brushes list. Select this new brush and click once in the file space. Change the size of the brush in the options bar, or change the colour in the Colour palette. If you click and drag you will get a repeat.

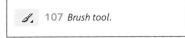
107 *Brush tool.*

108

108 *Brush tool options palette.*

109 *Background created using the Brush tool and a defined pattern, changing the colour intermittently.*

109

Regular repeats

A repeat pattern can be created by sampling an area of an image, ready to use as a repeat fill.

110 *Fill tool.*

Create a simple image such as a letter or a black-only shape, or open a shape that you have already created. Using the Rectangular Marquee tool, select the part of the image that you want. With the element still selected, Edit > Define Pattern from Selection. Name the pattern, keeping a logical naming system wherever possible. The samples section is now loaded in the Pattern Fill options box.

To apply, create a new file and select the Fill tool. In the Fill options bar across the top of your picture space, open the Fill drop-down box and select Pattern. Next, open the Pattern drop-down box and select the newly created pattern. Click with the Fill tool in the picture space and the area will be flooded with a repeat of the sample you created.

111 *Fill patterns.*

112 *Fill Options palette.*

113 *Background created with a pre-set Pattern Fill.*

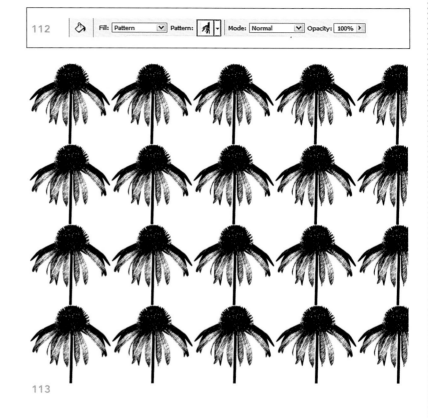

113

Cookie Cutters

Some versions of Photoshop have a Cookie Cutter tool. This is a simple tool to crop an image in a large range of pre-formatted shapes.

The process is no more difficult than cropping an image using the Crop tool.

- Open an image.

- Select the Cookie Cutter tool from the Tools palette.

Because of the images used to create the cutter shapes, another popular effect is available as a one-click process.

114 *Cookie Cutter tool.*

115 *Cookie Cutter options palette.*

There are several frame image options, some of which have a distressed edge. Although once you have gained experience creating images and effects like those discussed in this book you will be able to create your own frames, the Cookie Cutter gives plenty of effects to start with.

- Open an image.

- Select the Cookie Cutter tool from the Tools palette.

- Click the mouse in the image space and drag – like using the Crop tool.

- When the mouse is released the shape is automatically cropped. Press the return key to confirm the action.

Using the mouse to select an area either for cutting or any other action is usually carried out by eye, probably relative to the image. However in order to create a square, hold the shift key down while dragging the mouse and this will constrain the proportions.

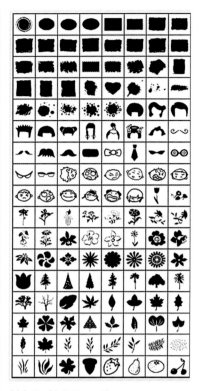

116 *Cookie Cutter options.*

117

117 *Cookie Cutter frame.*

Taking things further

13

Digital print bureaux

The type of printers listed below are not considered to be for small-scale or domestic use, and are described here as a comparison in case you are looking to upgrade, and to clear up any confusion when these types of printers are referred to. When you have developed your ideas and a desktop printer may no longer be your printer of choice, the printers listed here are available via a digital print bureau and prints can be individually commissioned.

Wide-format inkjet printers

If you choose to purchase a wide-format (industrial) printer or send your images to a digital bureau to be printed there will be more options available to you, and although not discussed extensively in this book, a summary of these is below. This is also an opportunity to clear up some confusion regarding fabrics that are identified as being pre-treated for digital printing.

Fibre-reactive and acid inks

Wide-format printers use fibre-reactive and acid dyes, and although they deliver the ink using an inkjet system, this will need to be fixed. In the normal use of these dyes, the dyed cloth is treated with mordants as part of the dyeing process to fix the dye. However if the dye is delivered by a printer, the cloth needs to be pre-treated with fixatives and activated with steam after printing to fix the dye.

Sublimation dyes (transfer and disperse dyes)

Sublimation dyes are manufactured specifically for inkjet printing by specialist companies (see Suppliers section on page 108) for a particular type of inkjet printers that use a non-thermal delivery system.

These inks, however, cannot be used in a regular inkjet printer, because the delivery system for delivering normal printer ink uses heat. Printers designed for using sublimation dyes require a non-thermal system. Sublimation systems are mentioned here as they are an option for you to explore; however, sublimation dye needs to be printed onto a specialist transfer paper to offset-print onto fabric or another substrate. Industrially printed synthetic fabrics are produced in this way, by printing the image or pattern from a wide-format printer onto wide paper and then heat transferring, all in one pass.

118 *The print of the face used in this piece is a simple black-only image but placed behind a piece of Evolon (synthetic fabric) and cut with a craft soldering iron to create a 3D effect.*

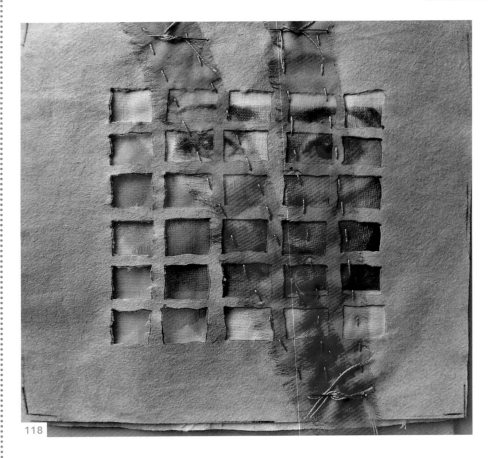

118

Collecting images

14

Sourcing images

Direct sources

In many ways, sourcing images is quite straightforward. If you are collecting and recording images from direct sources, such as going out with a camera and clicking away, it is often as simple as keeping your eyes open. But, like most things, it is not always quite as easy as that.

You need to develop a 'creative eye' or, put more simply, look at the world in a different way. You should consider surfaces, textures, lines and colour as separate elements. A good trick when looking at an object is to half close your eyes to create a different view. Even if you are looking at a photograph you have already taken, digital technology allows you to isolate parts of the image by cropping or selecting parts to create a brand new image. This is where making sure that your digital files are saved at a minimum of 300ppi comes into its own.

Indirect sources

A word of caution: if collecting from indirect sources such as images in books, documents or from the web, be aware of contravening the copyright rights of artists. However, it is generally acceptable to create single copies of images for personal use to record ideas. If downloading images from the web they have their own inbuilt 'security' as web images will only be formatted at 72ppi and when resized will break up and become pixelated.

Likewise, images scanned from a commercially printed source have their own inbuilt defence. Commercially printed images are created from a series of dots that cannot be seen by the naked eye, but if these images

▶ **HINTS & TIPS**

- Photograph flower heads rather than the whole plant to create impact.
- Focus in closely on your subject to record detail.
- Photograph the same subject at different times of the day or year.
- Look up! We don't look up often enough. It can be a different world.
- Photograph reflections. It isn't as difficult as you might think.
- Photograph mundane subjects, looking closely at their detail and colour.

are scanned, particularly at a minimum of 300ppi, the image will appear as if it is printed on a mesh. Better scanners will have a facility called 'descreening' which with some skill can be employed to counter some of this effect.

The design process

Think about copying your own images using an office photocopier; very often interesting, high-contrast results will appear giving you ideas of line and shape, without you really having to try too hard. The luxury of having access to an office photocopier means that images can be resized endlessly until only texture remains. The rule does not apply in quite the same way if printing on a colour copier, but interesting effects can still be achieved.

Having a camera with a macro facility will enable images to be captured at very close range, isolating areas of an object at source. A macro lens is sometimes confused with a telephoto lens, which does not do the same job at all. A telephoto lens enables subjects to be photographed from a distance.

As well as giving digital files a coherent file-naming system, try to apply this working practice to printed images you may have, by filing or grouping them according to colour, detail, texture, black and white or monotone.

Tear up an image and re-copy or re-scan it
A lack of experience will stop you from seeing an image as a sequence of colour and shape etc., but with more practice the more you will let go. *There are no rules.*

Working in both the digital and the real world demands a certain kind of discipline, especially if you cross-fertilise ideas. Printing out digital images and rescanning them can be a fruitful method of creating new images, much like increasing the scale of a print on an office copier. However, you must keep a filing system of image sources and files, or you may produce a very interesting image but be unable to repeat the process, as you will have no idea how the constituent pieces were created.

As with this book, there is no one particular linear design sequence to follow. You can jump off at any point in the process.

119 *Images collected from catalogues and postcards are rarely revisited. Take a page at random and photocopy it.*

Recording ideas

The term 'sketchbook' is used quite liberally by artists but it can often lead to misunderstandings for those new to the creative process. Sketchbook is mostly used as a term to describe a place where information (of any kind) is kept, rather than a place that contains watercolours or sketches used as preparation for large paintings, as it is traditionally understood.

Size and scale

One way to create interesting pieces of work is to experiment with the scale of the marks that you make. This is almost impossible using most media, such as a brush or pencil, but one unique aspect of working with digital images is that this can be done easily with the click of a mouse. Experimenting with scale is one of the simplest ways of creating an abstract or impressionistic image.

120 *Collages from sketchbooks.*

121 Helenium.

122

122 *One of many sketchbooks. A sketchbook is a place where information of any kind can be kept, rather than a traditional book of sketches.*

Reflections and mirrors

Sometimes you need to stand back and try to see the images that your brain is filtering out. The modern world is full of reflections, particularly in contemporary urban landscapes, but our brains will very often 'edit' them out. Stop for a moment and bring them back into focus.

Developing designs and ideas

Take time to print out your images in different sizes. Looking at images on a screen is a completely different experience to looking at a print. Colours become more subdued and subtle because much less light is being reflected. The subject matter will retreat further into its surroundings.

Print out images and use them for a paper collage. Combine these with different pieces of paper or print to create new images and or backgrounds. Rescan the paper pieces.

Line and shape

Get into the habit of taking photographs of everyday objects. Again, these are elements of the everyday world that you may look at but not really see because your brain 'edits' them out.

One thing that a camera is particularly skilled at is identifying and selecting the formal elements of an image or object, particularly if converted into grayscale.

Isolate individual elements such as line and shape by converting images to grayscale or bitmap images.

Texture

Placing images of similar subjects in different sizes or scales together can help to highlight differing textures, particularly if the image is displayed in full colour. Some of the images shown opposite have been edited to restrict the amount and type of colour information in order to emphasise texture.

Viewpoint

When we are out and about most of us see the world from around four or five feet above ground level. This means that many everyday domestic objects are seen from above. Get into the habit of going down to the same level as an object. It is amazing how a composition of objects suddenly becomes more interesting.

123

123 *This image was taken inside the Burrell Collection, Glasgow, where I felt that the building was an integral part of the exhibition.*

124 *These images were taken casually on a day out at an exhibition and in a shopping arcade.*

125 *Everyday, familiar objects and places look very different when viewed from a different angle.*

One of the big advantages of investing in a slightly more sophisticated camera with a macro facility is that objects can be recorded in extreme close up, whilst retaining all of the detail.

All of these techniques can create spectacular images without even going anywhere near photo imaging and digital enhancements.

Take a photograph of everything (within copyright and privacy of course!).

126 *A collection of images that have been converted to grayscale to highlight shapes.*

127 *A collection of images that have been converted to bitmap to highlight line and contrast.*

128 A group of images from the garden in full colour. No colours have been edited.

129 The same group of images have had a single colour changed to blue, whilst retaining a neutral background.

130 All colour has been removed by converting to grayscale, to highlight a certain amount of texture.

131 Grayscale images converted to a single colour to further highlight overall texture.

Converting from a photograph to an abstracted image

132

133

Everyday minutiae
The most interesting images live in the everyday.

Drawn images
When sketching, try to keep the image 'clean' by using an ink pen rather than a pencil as this will cause the image to break up when scanned. If the image is scanned in grayscale, your sketchbook paper will be visible within the image, so try experimenting with scanning the image directly as a bitmap or try converting it to a bitmap after scanning. Also remember to scan at a very high resolution, at a minimum of 600ppi.

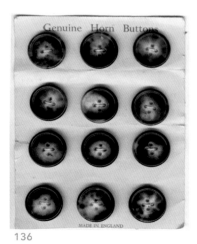

136

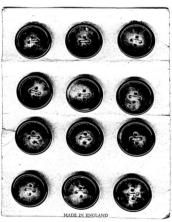

137

Some of the images used to illustrate techniques contained in this book are available to download at wendycotterill.wordpress.com

132 A collection of everyday objects.

133 The same objects have been converted to imitate a hand-drawn picture using photo-editing software.

134 The best carrot cake in the world!

135 This is a page scanned directly from a sketchbook as a bitmap.

136 Everyday objects scanned at high resolution to caption all of the detail. Notice that the white card background is captured as a darker background shadow.

137 When this image is converted to a bitmap, the background disappears. The image has not been 'cleaned up', i.e. by removing background speckles, as this adds to the overall antiqued effect.

Gallery

15

138

139

138 *A collage created in Photoshop using transparency in layers.*

139 *These very small pieces of bright white cotton fabric have been printed with simple text to create a background texture. Try using text randomly scanned from a book as the source of your text, rather than typing directly into an image file.*

140 (detail) A collage of text printed using the paint and organza surface and pieced together with smaller jewel-like images with hand stitching.

141 This image was created by following the scrim and PVA method of printing, which gives a textured surface.

142 *There are many elements to this piece of work as it is a mixture of many different fabrics and papers. However, the text was created by printing direct to fabric before the collage was assembled.*

143 *These butterflies were simply printed using a disk of digital images and then partially cut with a craft soldering iron.*

60 Astrological sy[...]
diagram. On the h[...]
the zodiac, assist[...]
interaction refines 60 Astrolo[...]
stages of birth[...] diagram. [...]
(Mo[...]tain of A[...] he zodiac
Augsburg, 1[...]) [...] nteraction

61 Enclo[...] by the ages of bi[...]
sun, moon and plan[...] Mountain
active, passive and u[...]gsburg,
creation. This print [...] Enclose
but the teaching of[...], moon
which abide in the b[...] ive, pass[...]
62 This table illustra[...]ation. T[...]
144 the metals and [...] the tea[...]

144 *Simple black text on a bright white background used as background texture.*

145 *The centre image used in this collage was printed onto synthetic organza and water allowed to bleed the image. If you try this technique you will need much more water than you might imagine.*

145

146 *This composite piece uses in part the scrim and PVA method described in Chapter 8, but also uses water to discharge some of the background image on paper, as well as additional stitching.*

147–149 *These images are striking on their own but they are actually cropped, close-up details of the image above, showing how cropping techniques can highlight texture and colour.*

146

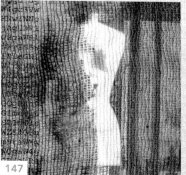

147

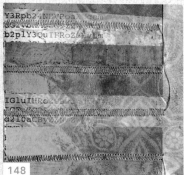

148

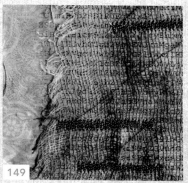

149

150

151

150 *Close-up detail of a collaged piece of text printed on organza with small embellishments.*

151 *This image was created by following the scrim and PVA method of printing, which gives a textured surface.*

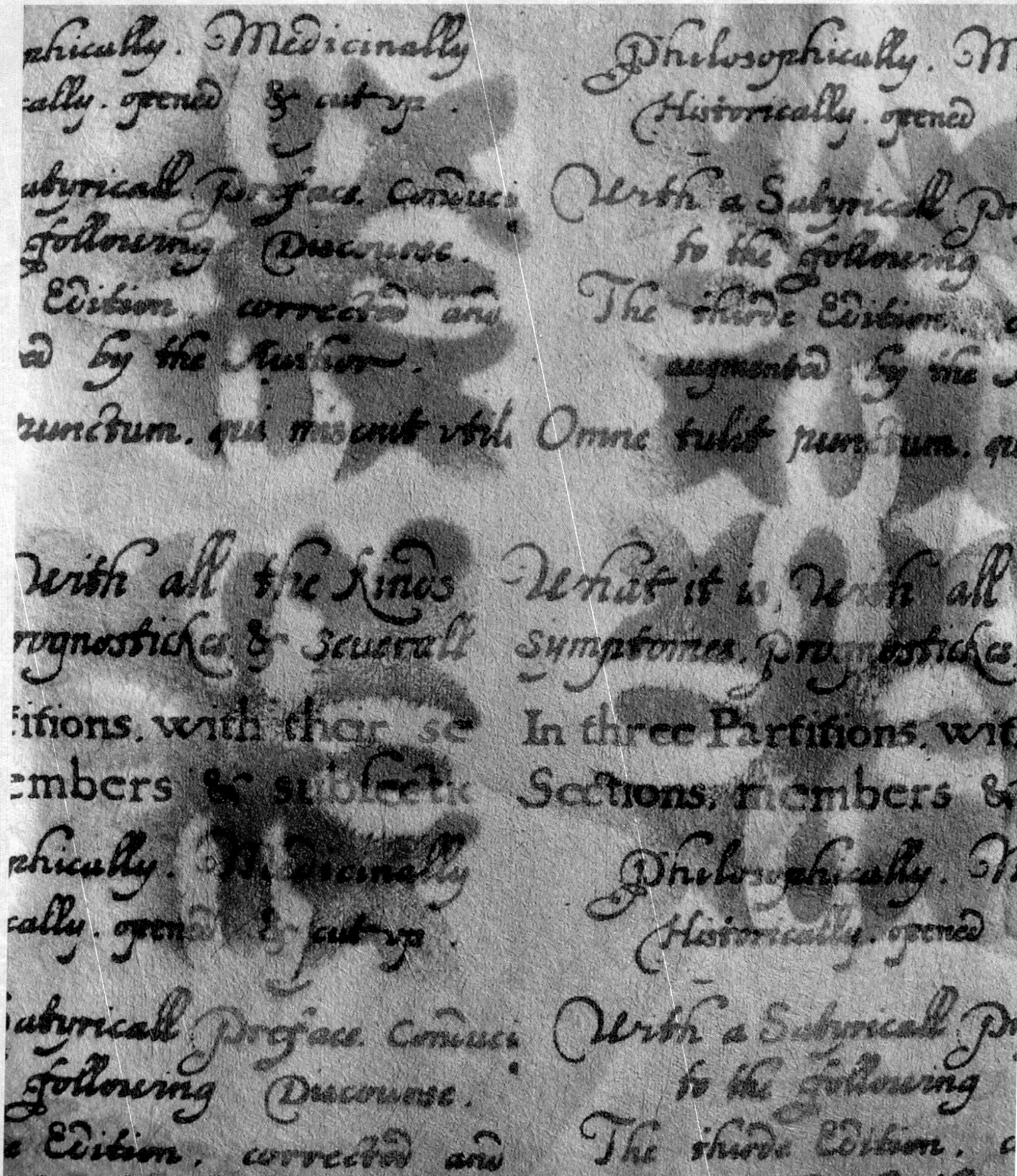

153

152 Black and Blue Text - *text printed onto Evolon, foam stamp sprayed with Print.Ability and over-sprayed with dye-based sprays.*

153 *Many different pieces of paper, fabric and scrim were collaged together to create this overall ethereal effect.*

Suppliers

UK

General supplies

Gallery Textiles
+44 (0)7906 971062
www.gallerytextiles.co.uk
*Print.Ability, Evolon, Lutradur,
Zeelon, craft soldering irons,
scrim, organza*

Rainbow Silks
6 Wheelers Yard
High Street
Great Missenden
Bucks
HP16 0AL
+44 (0)1494 862111
www.rainbowsilks.co.uk
*Bubble Jet Set, digital print
mediums*

Fabric

Whaleys (Bradford) LTD
Harris Court
Great Horton
Bradford
BD7 4EQ
+44 (0)1274 576718
www.whaleys-bradford.ltd.uk
*Digital fabric and all kinds of
loomstate and undyed fabrics*

Empress Mills
Glyde Works
Byron Rd
Colne
BB8 0BQ
+44 (0)1282 863181
www.empressmills.co.uk
*Good quality high thread count
plain cotton*

Digital bureaux

Fingerprint Fabric
135 Post Office Road
Seisdon
Staffs
WV5 7HA
+44 (0)07877 402455
www.fingerprintfabric.com
*Laura Kemshall prints fabric on
request using wide format printers
(contact for pricing)*

The Silk Bureau
Corn Mill Bank
Hinton on the Green
Evesham
WR11 2QU
+44 (0)1386 861122
www.thesilkbureau.co.uk

Magic Textiles Ltd
Unit 1A
Churnet Works
James Brindley Rd
Leek
Staffordshire
ST13 8YH
+44 (0)800 612 2801
www.magictextiles.co.uk

Netherlands

Zijdelings
Kapelstraat 93A
5046 CL Tilburg
Netherlands
31 (0)13 545 3456
www.zijdelings.eu
*European retailer of Print.Ability
and also runs classes in creative
textiles*

USA

Meinktoy Fiber Art Supplies
00 1 561 357 8300
www.meinketoy.com
*Retailer of Print.Ability in the USA
and supplies many of the creative
materials for creative textiles*

Australia

The Thread Studio
6 Smith Street
Perth
Western Australia 6000
+61 8 9227 0254
www.thethreadstudio.com
*Carries most of the creative
textiles supplies in Australia*

Further reading

Berger, John, *Ways of Seeing*, Penguin Classics, 2008

Bowles, Melanie and Isaac, Ceri, *Digital Textile Design*, Laurence King, 2012

Cole, Drusilla, *The Pattern Sourcebook: A Century of Surface Design*, Laurence King, 2009

Cotterill, Wendy, *Spunbonded Textile & Stitch*, Batsford Books, 2011

Dunnewold, Jane, *Art Cloth: A Guide to Surface Design for Fabric*, Interweave Press, 2013

Fish, June, *Designing and Printing Textiles*, Crowood Press Ltd, 2005

Hansen, Gloria, *Digital Essentials*, Electric Quilt Company, 2009

Steed, Josephine and Stevenson, Frances, *Basics Textile Design 01: Sourcing Ideas: Researching Colour, Surface, Structure, Texture and Pattern*, AVA Publishing, 2012

Links

Adobe Adobe Systems Software
www.adobe.com/uk
Computer Textile Design Group
www.ctdg.org.uk

Serif (Europe) Ltd
The Software Centre
12 Wilford Ind Est
Nottingham
NG11 7EP
www.serif.com

The Textile Artist
www.textileartist.org

Glossary

Bitmap A bitmap image is made up of a rectangular grid, or raster, of pixels, very much like a mosaic.

Bridge camera A camera that has some of the facilities of an SLR camera, combined with some of the automatic facilities of a compact camera.

CMYK (Cyan, Magenta, Yellow & Black) Images for use in a commercially printed publication will need to be created as, or converted to, an image that has the four colour channels: Cyan, Magenta, Yellow and Black.

Compact camera A digital camera that usually only has limited facilities in terms of lens capability, image resolution etc. but is normally better than a phone camera.

Compression artefacts As described below, .JPG is a compression file format and when a .JPG image is resized, the pixels contained within the image have to be combined or added to, to complete the resizing. If the file is then re-saved more than once, this pixel reconfiguration creates blotchy areas within the image, called compression artefacts.

Compression (lossy) In order to keep files small and use less disk space, compression formats delete any non-essential information each time an image is saved, which means that the quality of files saved in compression formats will decrease every time they are saved.

Compression (non lossy) Images can be compressed when saved without losing any of the original file information.

Descreening A commercially printed image is made up of small dots. Scanning can emphasize these dots, which then creates a pattern across the scanned image. Activating a descreening facility in scanning software can help to reduce or eliminate this.

Discharge This refers specifically to a process that printers and dyers use in which colour is removed from fabric. There are several ways that this could happen: there are specific processes that rely on the removal of colour from a background to create a pattern; or this can also refer to loose dye that runs off into water when newly dyed fabric is rinsed.

Dots per inch This is the measure of how much ink is printed onto a surface. It is often used interchangeably with the term 'pixels per inch' but the two can be very different.

File format When a digital file is saved it will need to be saved in a particular format, such as .JPG, .TIF etc. Images use different file formats from text-based documents.

JPEG (Joint Photographic Experts Group) .JPGs were designed to save space by deleting all non-essential information from a file each time it is saved. This leads to gradual image degradation, which is not reversible. It is recommended only to use .JPGs when creating images for the web. If this is the only method available to save a file, create a new file each time an image is edited. This is a lossy compression format.

Non-compression This format maintains all original file information every time a file is saved.

Non-polymer This is a term used in this book to refer to print mediums that are not derived from polymer technology.

Overprinting An image printed over a previously printed image.

Pixel A digital image is formed of many thousands of pixels, which when seen in magnification are simply tiny squares of colour. When reduced to normal magnification, the squares dither so that the eye only sees the overall image.

Pixels per inch Image resolution is measured by the number of pixels per square inch. This term is used interchangeably with 'dots per inch' (dpi) but does not accurately translate as the same measurement. DPI is a printer's term because when an image is commercially printed, the printer makes a series of tiny dots that dither to create a perfect image at normal magnification. When working with digital images, ppi is the correct measurement to use.

Polymer Polymer is a chemical term that refers to the molecular structure of many synthetic fabrics. These tend to be non-porous. Not to be confused with the term 'man-made' as this contains a group of fibres regenerated from natural sources, e.g. viscose is a man-made fibre derived from wood pulp.

Polymers can also refer to paint and its auxiliaries. Polymer-based print mediums are emulsions that create a waterproof surface.

PSD PSD is a software-specific file format and can usually only be opened within Adobe Photoshop. Other photo-editing programs will have their own specific file format. .PSD files will maintain layers and editable text. This format is a non-compression file format.

PVA (Poly Vinyl Acetate) A white glue originally developed as a water-based wood glue. It can now be purchased in many different varieties and strengths. The PVA glue referred to in this book is the original full-strength glue.

Rasterise To convert an image from vector to bitmap format. Cameras and scanners will capture an image or photograph as a bitmap image.

Resampling An automatic process by which the number of pixels in an image is increased or decreased when the image is resized.

Resizing Changing the dimensions (print size) of an image.

Resolution Resolution of a bitmap image is measured in pixels per inch.

RGB (Red, Green, Blue) An image saved to be seen primarily on a computer screen is made up separately of three colour channels: Red, Green and Blue. When this type of image is printed on a desktop printer, the printer software will convert the information from the onscreen RGB image into CMYK.

SLR camera (Single Lens Reflex camera) The image to be photographed is seen directly through the lens. Traditional manual SLR cameras use 35mm film. Digital SLR cameras have facilities to maintain information in an image to imitate the detail that is captured on a 35mm film.

TIFF (Tagged Image File Format) This is a non-software-specific file format for saving image files. You will be able to open it using most photo-editing software, but because it contains all of the file information needed to reconstruct the image every time it is opened, the file size will be unusually large. TIFFs will maintain layers and editable text. This is a non-compression format.

Vector An image made from a series of mathematical geometric objects. Vector images are able to be resized without any loss in image quality.

Index